Patrick M

Practical . HWLCSC

Other titles in this series

The Observational Amateur Astronomer
Patrick Moore (Ed.)

Telescopes and Techniques
C.R. Kitchin

The Art and Science of CCD Astronomy
David Ratledge (Ed.)

The Observer's Year
Patrick Moore

Seeing Stars
Chris Kitchin and Robert W. Forrest

Photo-guide to the Constellations
Chris Kitchin

The Sun in Eclipse
Michael Maunder and Patrick Moore

Software and Data for Practical Astronomers
David Ratledge

Amateur Telescope Making
Stephen F. Tonkin

Observing Meteors, Comets, Supernovae and other
Transient Phenomena
Neil Bone

Astronomical Equipment for Amateurs
Martin Mobberley

Transit: When Planets Cross the Sun
Michael Maunder and Patrick Moore

Practical Astrophotography
Jeffrey R. Charles

Observing the Moon
Peter Wlasuk

Deep-Sky Observing
Steve Coe

AstroFAQs

Questions Amateur Astronomers Frequently Ask

Stephen F. Tonkin

With 20 Figures

Springer

Patrick Moore's Practical Astronomy Series ISSN 1431-9756

ISBN 1-85233-272-7 Springer-Verlag London Berlin Heidelberg

British Library Cataloguing in Publication Data
A catalogue record for this book is available from the British Library

Typeset by EXPO Holdings, Malaysia
Printed and bound at the Cromwell Press, Trowbridge, Wiltshire
58/3830-543210 Printed on acid-free paper SPIN 10755568

For Louise, Tim and Siân, to whom I was a stranger whilst I was preparing this book.

Preface

An increasing number of people are taking advantage of the relatively low prices of astronomical equipment. Many of these people are doing so with little knowledge of practical astronomy and, as the volume of questions asked at astronomical society meetings and on internet newsgroups attest, there is a need for these questions to be answered in one place. Hence this book.

The fundamental premise behind *AstroFAQs* is that the beginning amateur astronomer wishes to get "up and running" with the minimum delay. A secondary premise is that anyone will better appreciate why something is done as it is if there is an understanding of the underlying principles. *AstroFAQs* addresses both these premises.

AstroFAQs makes no pretence to go into great depth – that would be impossible in such a slim volume – but it will give you the kick-start you need to choose and use your instrument effectively, and will take you to a level of expertise that is significantly higher than the "beginner" status. It uses a hierarchical section numbering system that simplifies cross-referencing. Suggestions for more in-depth reading are given throughout.

More and more "newbie" astronomers are entering this fascinating hobby by purchasing one of the "gee-whiz" *GOTO* telescopes, of which there is an increasing selection. These serve the wish to begin observation as soon as possible, but they do so at a price premium. With very little extra effort it is entirely possible to gain as much, if not more, joy and satisfaction from astronomy from a much simpler, and hence much less-expensive, instrument. Whilst *AstroFAQs* addresses the requirements of those who choose this latter approach, the tips and techniques are applicable to all telescope users.

Stephen Tonkin
Alderholt
May 2000

Contents

Chapter 1

Choosing Equipment

1.1 Choosing Binoculars

1.1.1 What do the numbers mean?

Two numbers, e.g. 10×50 or 11×80, designate a binocular. The first of these numbers is the magnification; the second is the diameter of the objective lenses in millimetres. For example, a 7×50 binocular has a magnification of $\times 7$ and objective lenses of diameter 50 mm.

From these numbers, you can calculate the size of the exit pupil by dividing the objective diameter by the magnification. For example, a 20×80 binocular has an exit pupil of 80 mm/20 = 4 mm. The brightest images from a telescope of a particular aperture will be obtained if the exit pupil is no smaller than the dark-adapted pupil of your eye is; i.e. about 7 mm in those under 25 yrs, declining to about 5 mm by the age of 50.

1.1.2 What is the best binocular for astronomy?

Good general-purpose instruments that are hand-holdable and excellent for astronomy are 7×50 or $10 \times$

50 binoculars, although even the most humble binocular will reveal a wealth of detail invisible to the naked eye.

Larger binoculars, such as 15×70, 11×80, 20×100, etc., allow you to see much more but they will need to be mounted on a tripod or other mount; and smaller ones give less-bright images, although they can have the advantage of being extremely portable.

For bright images, try to get a binocular whose exit pupils match those of your eyes.

There are some things to consider when buying a binocular for astronomy:

- Is the instrument soundly made?
- Lightweight binoculars are less tiring to hold than heavier ones.
- Quick-focus ones are difficult to focus sharply and they are also quick-defocus!
- Zoom binoculars rarely give images of as good quality as normal ones.
- Fixed focus binoculars are not suitable for astronomy.
- Don't get the high magnification chain-store ones, which are sometimes sold as good for astronomy – they are not! The images are not bright and they are difficult to hold steadily.
- Hold it at arm's length away from you and check that the circles of light you see are truly circular. If they have flat edges, the binocular is made with undersized prisms which will not pass all the light gathered by the "big end" to your eye.
- Take it outside and look at something like a distant television aerial against a light sky. Are there distracting coloured fringes? Does the image get very blurred at the edge of the field of view?
- Are the lenses coated? Try to get a Fully Multicoated binocular.

You will find that there is a great deal of difference in price, even in binoculars of the same size. Get the best you can afford; they will tend to have fewer faults and give better images. However, the best thing to do is to try some out first. If your local astronomical society holds observing evenings/field meetings, go along and try as many binoculars as you can and solicit as much advice as you can. If you have friends with binoculars, see if any of them suit you.

1.1.3 How should I hold a binocular?

This first answer is that you should *not* hold a binocular in the "normal" way, i.e. by the objective barrels. This, although seemingly natural, is not very stable.

A far better method is the "triangular arm brace" method. Hold the binocular with your first two fingers around the eyepieces and the other two fingers around the prism housing. Then raise the binocular to your eyes and place the first knuckle of your thumbs into the indentations on the outside of your eye sockets, so that your hands are held as if you were shielding your eyes from light from the side. Each arm is locked into a stable triangle with the head, neck and shoulder as the third "side", giving a relatively stable support for the binocular. The position of your thumbs keeps the eyepieces a fixed distance from your eyes. You cannot normally reach the focus wheel on centre-focus binoculars when you hold them this way, but you should not need to refocus during an observing session. This grip does feel unusual at first, but it is so superior to the "normal" way that it soon becomes second nature.

That is still not the most stable way. If you want the most stability you can get with a medium sized hand-held binocular, use the "rifle-sling" method. This is similar to the way one uses a sling for rifle-range shooting. Hold the binocular so that the strap loops down. Place both arms through the strap, so that it comes just above your elbows. Hold the binocular in the most comfortable way you can and brace it "solid" by pushing your elbows apart. It's a bit like getting into a medieval torture instrument, but it's very effective.

Lastly, you may sometimes wish to handhold a large binocular for short periods, and find that the balance of the binocular makes the "triangular arm brace" method unstable. Assuming your right eye is dominant, use the "triangular arm brace" with your right hand and hold the right objective barrel with your left hand. The left objective barrel is supported by your left wrist (you may wish to remove your wristwatch first). For extra stability, combine this with the "rifle sling" method.

1.1.4 How can I mount a binocular?

A mounted binocular will give a much clearer image and will allow the observation of objects a magnitude or more fainter than those observable in the same binocular when it is hand-held. The simplest way to mount a binocular is using an L-bracket that attaches the binocular to a camera tripod (see Figure 1.1).

This is satisfactory except when observing objects near the zenith, when the observer's limbs tend to become entangled with those of the tripod.

More satisfactory are parallelogram mounts that are counterweighted and hold the binocular away from the tripod (see Figure 1.2). They have the advantage that the binocular can be raised or lowered to different eye heights, whilst maintaining alignment on the target object. This is very useful when adults and children are sharing a binocular.

Figure 1.1 A binocular mounted on a tripod L-Bracket

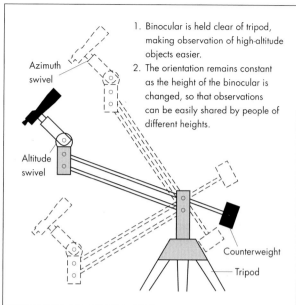

1. Binocular is held clear of tripod, making observation of high-altitude objects easier.
2. The orientation remains constant as the height of the binocular is changed, so that observations can be easily shared by people of different heights.

Azimuth swivel

Altitude swivel

Counterweight

Tripod

Figure 1.2
Parallelogram Binocular Mount

1.1.5 Where can I find out more?

Patrick Moore's *Exploring the Night Sky with Binoculars* is an excellent introduction to binocular astronomy.

1.2 Choosing a Telescope

1.2.1 Do I need to buy a telescope?

Assuming you are new to amateur astronomy, the answer is probably "No, not yet". You are probably better off with a decent binocular (see Question 1.1.2). It would certainly be advisable to try out a few telescopes (e.g. at a field meeting or star party) before you commit yourself to a purchase. Astronomical societies often have field meetings where newcomers can use members' instruments under guidance, and the societies themselves are usually sources of copious free advice.

1.2.2 Where should I buy a telescope?

An astronomical telescope is a specialist scientific instrument and should therefore be bought from someone who can give you appropriate and reliable pre-sale advice, and good after-sale service. The obvious choice is then a reputable astronomical telescope retailer. The obvious non-choice is a department store! You should be able to get recommendations as to reputable suppliers from your local astronomical society.

Second-hand telescopes are obtainable through classified advertisements in the astronomical and other press, and at auction. You need to know what you are looking for and, if you do not have the requisite experience, enlist a friend who does.

1.2.3 How do I decide upon a telescope?

There is an increasingly large range of telescopes available and it is easy to be overwhelmed by the choice. You should list the things you wish to be able to do with your instrument and then prioritise that wish-list. Some things you may wish to consider are:

- Portability. If the telescope will need to be moved around a lot, e.g. from an urban home to a dark observing site, you are more likely to do this if it is not a severe test of your physical strength. Your instrument should also easily fit into whatever transport you will use – tight fits are no fun in the dark; neither is manoeuvring heavy optical tubes on and off a mount!

- Ease of set-up. A telescope which is easy to set up will get much more use than one that involves a great deal of time and effort – this is one reason why Dobsonian-mounted telescopes are so popular.

- High-magnification planetary observing. You will either need an equatorial mount or a computerised altazimuth mount – tracking smoothly enough to make detailed observations at high magnification is a difficult art to learn with a manually-controlled altazimuth mount.

- Astrophotography. Unless you make only very short exposures, you will either need a motorised equatorial mount or a computerised altazimuth mount with field de-rotator. If you will be doing prime focus, afocal, or eyepiece projection photography you will also need some form of drive corrector and a guidescope (or off-axis guider) with an illuminated reticle eyepiece.

- Accessories. You may like to consider what standard accessories are available for your telescope. Some, like eyepiece filters or drawtube T-ring adaptors, can normally be considered to be "generic". Others, like coma correctors or motor drives, are usually specific to a particular instrument or a small selection of instruments.

- Upgradeability. Unless you are fortunate enough to be able to buy your "dream system" in one go, you may wish to buy a basic outfit to begin with, and then upgrade it as and when resources allow. If this is the case, you would be wise to investigate how easy it would be to upgrade. For example, would it be possible to remount the OTA of your budget Dobsonian onto a good equatorial? How easy is it to add drives or digital setting circles to your mount?

- Refractor, reflector or catadioptric? There are many opinions offered on this topic, but much of it is based on prejudice and tradition. These are the salient facts:

    Reflectors give you the greatest aperture for your money, but tend to go out of collimation easily and can suffer from tube currents. The mirrors also need periodic recoating. They are justifiably favoured by deep sky observers on the grounds that "aperture is king" when the target is often at the limit of visibility. They have the advantage that the only source of chromatic aberration is the eyepiece. Short focal length reflectors are very demanding of eyepieces and require them to be of excellent quality. Reflectors are usually also more amenable to tinkering, which can be a boon if you wish to perform DIY improvements.

    Refractors tend to require less frequent recollimating and do not suffer from tube currents. Planetary observers traditionally favour them on the grounds that the image is not degraded by the central obstruction of a reflector or catadioptric, but an optimised planetary reflector (e.g. a long-focus Newtonian) can give as crisp an image. All refractors will lose some light into the secondary spectrum as a consequence of chromatic aberration, but this is minimal with modern top-end designs. Good refractors are very expensive.

Catadioptrics tend to be very sensitive to miscollimation. Cassegrain-type systems (including Maksutov-Cassegrains and the ubiquitous Schmidt- Cassegrains) are more compact, but all have a large central obstruction. They are a good compromise as a relatively portable, general-purpose option. They should never be used for solar projection because of the potential damage that can be caused by the inevitable internal build-up of heat. Schmidt-Newtonians and Maksutov-Newtonians should be considered as specialised instruments and are usually priced accordingly.

1.2.4 What does "diffraction-limited" mean?

Strictly speaking, "diffraction limited" means that any aberrations in the telescope's optics are sufficiently small that they render the optical system indistinguishable from a perfect system, where the limits are those imposed by diffraction and are related to the aperture of the instrument. There is no universally accepted definition and, in reality, the phrase means whatever a manufacturer decides it should mean. Therefore, identical claims in this regard made by different manufacturers may mean different things. Manufacturers began to use the term when their numerical wave ratings came under scrutiny.

1.2.5 What do manufacturers' wave ratings mean?

The short answer is "they are usually meaningless". The Rayleigh criterion is that an optical system will be indistinguishable from perfect if the wavefront error is less than $\lambda/4$, where λ is the wavelength of light. A reflective surface doubles any wavefront error, so the surface error should not exceed $\lambda/8$. A refractive surface halves the error, so its surface error could be as large as $\lambda/2$.

Manufacturers often claim that their telescopes have "eighth-wave" optics. This is meaningless unless they also tell you:

- Is this a peak-valley (PV) or a root mean square (RMS) rating? Although RMS is sometimes more appropriate to matters concerning waves, the Rayleigh criterion refers to PV values, and the PV error is 1.4 times as great as an RMS one. To satisfy the Rayleigh criterion, the RMS error of a reflective surface must not exceed $\lambda/11.2$.

- The wavelength of light to which the claim refers. It may refer to red light (wavelength in the region of 700 nm), but the eye is most sensitive in the yellow-green region of the visible spectrum (around 500 nm), i.e. that region for which refractor objectives are optimised. To satisfy the Rayleigh criterion at 500 nm, the error of a reflective surface at 700 nm must again not exceed $\lambda/11.2$.

- Does the claim refer to all parts of the optical system or just, in the case of a reflector, to the primary mirror? The image from an excellent primary could be ruined by a poor secondary.

- Does the claim refer to the entire surface? If this is not the case, any error is likely to be at the edge of the mirror, where it will have the most deleterious effect on the image – nearly 20% of the light falls on the outer 10% of the mirror's diameter!

- How was the error measured? The only reliable quantitative ratings are those obtained by interferometry. Most mass-manufacturers do not test every telescope with an interferometer; for some mass-producers the quality control is performed by the customer!

Even if all that is given, it can only give you an indication of the error at the wavefront.

1.2.6 Where can I get more information?

An excellent book for guiding your choice is: Harrington, P., *Star Ware*, 1998, John Wiley and Sons.

1.3 Choosing Eyepieces

1.3.1 How many eyepieces do I need?

The obvious minimum for visual observation is one. You will probably want at least two, and this should be considered to be a useful minimum. As for more, just remember that it is better to have a few good eyepieces than lots of poor ones.

1.3.2 What is meant by low-, medium- and high-power?

Low power: magnification of about a fifth to a quarter the aperture of the telescope in millimetres. The focal length of the eyepiece (in millimetres) will be approximately four or five times the focal ratio of the telescope.

Medium power: magnification about equal to the telescope aperture in millimetres. The focal length of the eyepiece (in millimetres) will be approximately the same as the focal ratio of the telescope.

High power: magnification about one and three quarters times the telescope aperture in millimetres. The focal length of the eyepiece (in millimetres) will be slightly more than half the focal ratio of the telescope.

Thus, for a 100 mm refractor, this would equate to magnifications of about ×25, ×100 and ×175. It is often said that the highest power a telescope can sustain is equal to double its aperture in millimetres (or ×50 per inch of aperture). This is only usable in ideal conditions – these do not usually apply.

1.3.3 What focal lengths give these powers?

You can find the required focal length, fe, using:

$$fe = D \times F/M$$

where D is the aperture, F is the focal ratio of the objective or primary, and M is the magnification.

Example: a 150 mm f/5 telescope to be used at a magnification of ×40 will require an eyepiece of focal length $= 150 \times 5/40$ mm
$= 18.75$ mm
The nearest common focal length is 20 mm, which will give a magnification of ×37.5

1.3.4 Which two eyepieces should my first two be?

Ones which give low and medium powers.

1.3.5 What types of eyepiece should I get?

The best you can afford. It is wise to try out various eyepieces before you buy – see if your local astronomical society has field meetings where you can do this by borrowing the eyepieces of other members.

For general-purpose eyepieces, Plossl and Orthoscopic are normally adequate. Erfles are good for low magnifications. If you are on a budget, you could consider a Kellner on a telescope with a focal ratio of about f/7 or greater. If your telescope is faster than about f/5, you will probably need a higher quality eyepiece such as a Nagler, a Radian, or a top-end Plossl.

1.3.6 Why do low (fast) focal ratios require higher quality eyepieces?

The angle of the light cone produced by the primary mirror or objective lens is purely a function of the focal ratio; smaller focal ratios produce more obtuse angles. A "fast" light cone therefore impinges on a greater area of the field lens of the eyepiece and does so at steeper angles than does a "slow" light cone. The field lens therefore has to be of excellent quality over a larger area.

You should also be aware that fast refractors make greater demands on a star diagonal because of the more obtuse angle of the light cone. If you use a prism diagonal, this obtuse angle is more likely to result in dispersion of the light and false colour in the image.

1.3.7 How does a Barlow lens work?

A Barlow lens (see Figure 1.3) is a diverging or "negative" lens that increases the effective focal ratio of an objective lens, thereby increasing the magnification. The principle is identical to that of the "teleconverter" lenses used with 35 mm SLR cameras. The idea is that two eyepieces and a Barlow will give you the same flexibility of magnification as will four eyepieces, and will give higher magnifications with less powerful eyepieces.

By increasing the focal ratio, the Barlow lens reduces the angle of the light cone entering the eyepiece. The

Figure 1.3 Barlow Lens

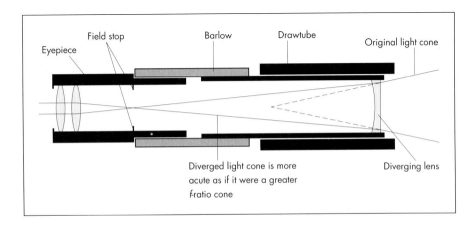

light cone therefore impinges upon the field lens of the eyepiece at a much smaller angle and over a smaller area; it is therefore much less demanding of eyepiece quality.

Other advantages of Barlow lenses are that they increase the eye-relief of the eyepiece and that they reduce off-axis aberrations like astigmatism and coma.

1.3.8 Will a Barlow lens work with all telescopes?

In principle, yes, but there is no guarantee that any Barlow will work with any telescope. The focuser of the telescope may have insufficient range for a particular Barlow, and a star diagonal can complicate their use. You should verify that the one you choose to purchase is compatible with your telescope.

1.3.9 Is there truth in the stories that Barlows give poor images and "ghost" images?

This certainly was true of some early Barlows, and still is the case with the singlet-lens Barlows that come with some budget telescopes. Modern, achromatic Barlow lenses with adequate anti-reflective coatings do not suffer from similar defects and have done much to dispel the bad name that Barlows have had in the past.

1.3.10 Where can I find out more?

Philip Harrington's *Star Ware* offers excellent advice.

1.4 Choosing Accessories

1.4.1 What filters are helpful for observation?

Broad Band Light Pollution Reduction (LPR) Filters

A broadband LPR filter is intended to block light emitted by street lighting, specifically sodium (high and low pressure) and mercury vapour. Obviously, they also block out light of the same wavelengths from astronomical objects. Those deep sky objects which are composed of stars (clusters, galaxies) or which reflect starlight (reflection nebulae) are generally not significantly improved by LPR filters, but some improvement may be seen in some emission nebulae, particularly planetary nebulae (e.g. M57) and supernovae remnants (e.g. the Veil Nebula). When you buy any LPR filter, you should ascertain what light it passes and how much it attenuates the "blocked" light. Some old LPR filters only block out the low-pressure sodium lines. An LPR filter will behave similarly to a blue filter on Jupiter.

Narrowband Filters

These are usually marketed as "Ultra High Contrast" (UHC) or "Ultrablock" filters. They are very good on planetary and emission nebulae (see Table 1.1).

Table 1.1. Summary of bandpass filter applicability

	Planetary and Emission Nebulae	Reflection Nebulae	Galaxies	Stars and Star Clusters
Broadband	**/***	*/**	*/**	*
Narrowband	****/*****	*	*	*
Line	*****	–	–	–

– = reduces visibility * = poor ** = moderate improvement *** = good **** = very good ***** = excellent

Line Filters

The two usually found are Oxygen-III (O-III) and Hydrogen Beta filters. O-III filters are remarkable on planetary and most emission nebulae. Hydrogen Beta filters have limited advantage over O-III filters, but a few nebulae (e.g. the Horsehead) do benefit from it. Line filters are very poor on objects that consist of starlight or reflected starlight (see Table 1.1).

Neutral Density (ND) Filters

Neutral Density filters are usually sold as "Moon Filters". They can reduce the amount of glare on the Moon and bright planets. They can also help split double stars, especially those where the components are of markedly different magnitude, by reducing the glare of the brighter component and reducing the intensity of its diffraction rings.

Polarising Filters

Used singly, a polarising filter can reduce the amount of reflected glare, rendering more detail visible on the Moon and planets. They can be used as ND filters with double stars. Used in pairs, by rotating one filter the effect is a variable ND filter which blocks anything from 70% to 95% of the incident light.

Colour Filters

Colour filters are most useful on solar system objects. Table 1.2 summarises their use.

1.4.2 What finder should I choose?

There is an ongoing debate about the relative merits of conventional and unit-power (sometimes erroneously called "zero power") finders. Each has its advantages.

Table 1.2. Applicability of Colour Filters

	Red	Orange	Yellow	Green	Blue	Violet
Sun		Use with mylar solar filter to restore colour				
Moon	Enhances features	Greatly enhances features	Enhances features	Enhances features	Enhances features	
Mercury	Reduces sky brightness during daylight and twilight observations	Reduces sky brightness during daylight and twilight observations			Reveals some surface markings at twilight	Enhances faint features
Venus	Use during daylight to reduce sky brightness	Use during daylight to reduce sky brightness	Reveals some surface features	Reveals cloud patterns	Reveals some cloud detail	Improves cloud contrast
Mars	Enhances polar caps	Darkens maria and canali	Darkens maria and canali	Increases contrast of polar caps and dust storms	Improves polar caps and some surface features	Improves polar caps
Jupiter		Enhances detail in belts and festoons	General enhancement of surface detail	Improves contrast of Great Red Spot	General enhancement of red belts and Great Red Spot	
Saturn		Enhances bands	General enhancement and clarifies Cassini's division		Enhances low-contrast features	Enhances detail in the rings
Uranus			Shows some detail in large telescopes (>250 mm)			
Neptune			Shows some detail in large telescopes (>250 mm)			
Comets		Enhances dust tails		Can reveal some structure in brighter comets	Increases detail in gas tails	Can reveal some detail in bright comets
Comments						Do not use on small telescopes (<150mm)

1.4.3 How does a unit-power finder work?

Unit-power finders (also known as reflex finders) are mini "head-up displays". They project the image of either a red point or of a series of concentric circles onto the sky and the telescope is aimed by placing the dot or the bull's-eye over the target.

1.4.4 What is the best unit-power finder?

By general consensus, those, such as the *Telrad*®, which project concentric circles of known angular size onto the sky, are more useful. They do, however, tend to be more prone to dewing. Some of the "red dot" finders have the virtue of extreme simplicity. Those with variable brightness and the facility for "blinking" are more useful.

All are very easy to use.

1.4.5 What is the best conventional finder?

Opinions differ. However, it should have an aperture of at least 50 mm and a magnification in the region of × 8 to × 10. They pass nearly three times as much light as 30 mm finders which often come as a standard accessory and which are, at best, adequate. The 24 mm finders that come with budget telescopes are inadequate, not least because they are often stopped down even further to about 10 mm in order to reduce the aberrations of the singlet objective lens.

1.4.6 Should I choose a right-angled finder or a straight-through one?

Right angled finders, used with Cassegrains and refractors will, if the main telescope is used with a star diagonal, have the same image orientation as the main telescope. Some people consider this to be helpful.

There are some straight-through finders available which have an internal roof-prism in order to give an uninverted image, i.e. of the same orientation as the naked-eye view. Some people find this to be helpful

Most finders are straight-through with an inverted image. They take some getting used to, but most people manage perfectly well with them. They tend to be less expensive than the other two types for equivalent optical quality.

1.4.7 Can I use a conventional finder as a unit-power one?

Yes, as long as it is a straight through one. In fact, the best way to use a conventional finder incorporates this. Use both eyes, so that your brain superimposes the image of one eye over that of the other. Swing the telescope to place the cross-hair on the target area, then close the unaided eye to fine-tune the pointing. This is even easier if the finder has an illuminated reticle.

1.4.8 What lighting should I use for charts, setting circles, etc.?

Conventional wisdom is that the only acceptable light is a dim red one. Dim red light does not affect dark-adaptation (see Question 4.1.10). There are a number of red light emitting diode (LED) torches available on the market; LED torches have the advantage of an extremely low current drain, so they are very economical with respect to battery use.

However, many people find it difficult to read by red light; this is particularly true of those with presbyopia. Such people may find a very dim green light to be preferable, although there is no clear consensus on its affect on dark-adaptation.

1.4.9 What is an illuminated reticle eyepiece?

An illuminated reticle eyepiece has an engraved reticle that is illuminated by a red LED. It is normally used in conjunction with a guidescope or off-axis guider for guided astrophotography or charge coupled device (CCD) imaging, although some reticle designs are intended for use with double-star work, where they are used to measure separation and position angles. For guiding, the brightness must be adjustable so that the reticle does not obliterate a faint guide star. Pulsating illumination is also available.

1.4.10 What is a guidescope?

A guidescope is any telescope used to guide astrophotography or CCD imaging. For piggyback photography the main telescope can be used for guiding. For photography through the telescope, an auxiliary telescope is firmly mounted to the main telescope. Once the target has been acquired in the main telescope, the guidescope is adjusted until its reticle centres on a star and you then use the mount's drive-corrector to hold it in position during the exposure. Guiding can also be achieved electronically (autoguiding) using a CCD on the guidescope; this CCD is directly connected to the telescope's drive electronics.

1.4.11 What is an off-axis guider?

An off-axis guider attaches between the telescope and a camera. A small mirror or prism diverts a small amount of light from the periphery of the light cone through 90°, where an illuminated reticle eyepiece can examine it. In this way, a guide star can be acquired and used for guided astrophotography. Their use avoids possible problems of flexure between a conventional guide-scope and the main telescope. Conventional guidescopes are generally able to cover a larger portion of the sky, making it easier to acquire a bright guide star.

1.5 Astrophotography

This section is intended merely to give an introduction to astrophotography, which is sufficient to get you started. If you want deeper insights into this fascinating subject, I strongly suggest you purchase a specialised book (see Questions 1.5.10 and 1.5.11).

1.5.1 What camera should I use?

For general use, the best camera is a manual 35 mm SLR camera with a "B" setting on the shutter. Useful features include:

- Manual shutter – you don't want batteries running out during a long exposure.
- Mirror lock-up – reduces vibration when the shutter is fired.
- Interchangeable focus screens – microprisms and split rings are not useful for astrophotography.
- Interchangeable lenses – this is standard on most SLRs.

The camera does not need a light meter.

1.5.2 How do I use the camera?

There are four common modes of astrophotography:

- Piggyback – the camera rides on the telescope or the mount and takes photographs through its own lenses.
- Prime focus – the camera is attached to the telescope so that its focal plane is at the prime focus (strictly speaking, the Newtonian or Cassegrainian focus) of the telescope, which it uses as a lens.
- Eyepiece projection – an eyepiece (or, less commonly, a Barlow lens is introduced into the optical train, in order to obtain a more highly magnified image.
- Afocal – the camera is used, with its lens, to photograph directly into the eyepiece.

1.5.3 How do I connect a camera to a telescope?

You need two accessories for this:

- A T-ring that matches the camera. This attaches to the body of the camera and contains an insert with a female T-thread.
- A camera adapter to suit the telescope. This will fit into either the drawtube of the focuser or onto the back of an SCT. It has a male T-thread on the other end which mates with the T-ring on the camera. It may have the facility to hold eyepieces for eyepiece projection photography.

1.5.4 What other accessories do I need?

For unguided astrophotography, the only other essential accessory is a cable-release for the camera's shutter. For guided photography you need a guide-scope or an off-axis guider; either will need a guiding eyepiece with an illuminated reticle.

1.5.5 How do I focus the camera?

It is possible to use the camera using the focus screen, although this is not always easy. It is simpler to acquire or make a Hartmann mask, which fits over the aperture of the camera or lens. The mask has two holes, giving two images of a bright star, which will merge as one when the telescope is focused at infinity. The holes should be about 50 mm diameter, widely spaced, and neither must be impeded by the telescope's central obstruction, if it has one.

1.5.6 What film should I use?

For your first foray into astrophotography, ISO 400 slide film will be acceptable for deep sky objects, and ISO 100 for solar system objects.

Table 1.3. Bright Object Exposures, ISO 100, f/8

Object	Shutter Speed (seconds)
Moon, Full	1/500
Moon, Gibbous	1/125
Moon, Quarter	1/60
Moon, Crescent	1/8 to 1/30
Mercury	1/125
Venus	1/2000
Mars	1/125
Jupiter	1/60
Saturn	1/15

1.5.7 What exposure should I use?

Table 1.3 gives suggested exposures using ISO 100 film at f/8 for bright solar system objects.

For faint deep sky objects, you will need to experiment. The limiting factors are likely to be skyglow, which will fog the film, and your ability to guide accurately.

You should always bracket your astrophotographic exposures. Bracketing is the taking of exposures a few f-stops either side of the one that is presumed to be ideal.

1.5.8 What is the "hat trick"?

Even with a mirror lock-up, a camera can vibrate when the shutter is fired. For long exposures, the simplest way of overcoming this problem is to expose as follows:

- Place a dark hat (or similar item) over, but not touching, the aperture.
- Fire the shutter with the cable release and ensure that it is locked open.
- Remove the hat and begin timing.
- At the end of the exposure, replace the hat (again, not touching the telescope or camera), and close the shutter.

1.5.9 What precautions should I take with refractors?

Achromatic refractors are normally optimised for the eye's maximum sensitivity, i.e. in the yellow/green region of the spectrum. Camera films (and CCDs) have different sensitivities; hence a yellow filter (e.g. Wratten #12) should be used when using a visual achromat for astrophotography. A #12 filter has a transmission of 74%, so exposures should be increased accordingly. This precaution need not be taken with apochromats or photo-visual objectives.

1.5.10 Where can I get more information?

The best book on the subject is: Covington, M., *Astrophotography for the Amateur*, 1991, Cambridge University Press.

1.5.11 What about CCD imaging?

CCD (charge coupled device) chips are a fraction of the size of a 35 mm film frame. Care therefore needs to be exercised when matching a CCD with a telescope. This is an involved subject that is beyond the scope of this book. For more information see: Berry, R., *Choosing and Using a CCD Camera*, 1992, Willmann-Bell.

Chapter 2

Setting Up

2.1 Polar Alignment

In order to track an object successfully with a single motion, the polar (RA) axis of an equatorial mount must be parallel to the axis of rotation of the Earth.

When it is tracking at the sidereal rate, the polar axis of the mount rotates so as to counteract the rotation of the Earth.

In what follows, it is assumed that the polar and declination axes are mutually perpendicular, and that the optical axis of the telescope and the declination axis are also mutually perpendicular. There are three levels of accuracy in polar alignment: the "rough method", the "intermediate method" and the "accurate method".

2.1.1 What is the "rough method"?

This is the simplest of all, but has the advantage of being rapid and is suitable for visual observing.

- Adjust the altitude of the polar axis so that its angle to the horizontal is equal to your latitude. Either a wedge (whose angle is that of your latitude) and spirit level, or a plumb-line and protractor, are useful for this if there are no suitable markings on the mount.

- Set up the mount so that the polar axis is pointing as nearly north as you can judge.
- Align the telescope parallel to the polar axis of the mount and clamp it in position.
- Sight through the telescope, adjusting the altitude and azimuth of the polar axis until Polaris is centred in the eyepiece.

There is little point in offsetting towards the celestial pole, since it is likely that the error induced by aligning to Polaris will be exceeded by error in judging when the telescope and polar axis are parallel. The telescope will be sufficiently well aligned for visual tracking for shortish periods, but objects will appear to drift in declination.

If you are setting up your telescope somewhere from where Polaris is not visible, you will need to do the following:

- Adjust the altitude of the polar axis so that its angle to the horizontal is equal to your latitude. Either a wedge (whose angle is that of your latitude) and spirit level, or a plumb-line and protractor, are useful for this if there are no suitable markings on the mount.
- Obtain the difference between magnetic north and true north for your location.
- Use a compass to set your polar axis to true north. Be careful not to let any iron in the mount affect your compass.
- You can also find true north by marking the shadow of a plumb line at true local noon (i.e. sundial noon, or Equation of Time applied to Local Mean Time).

2.1.2 What is the "intermediate method"?

This is suitable for mounts with setting circles or polar-alignment scopes, such as bore-scopes. A bore-scope is a small telescope, usually with an illuminated reticle, inside the polar axis of a German equatorial mount. Some other mounts have detachable sighting scopes. For all these, follow the manufacturers' instructions, as the exact method of set-up will be specific to the mount and the alignment scope. The general principle of these alignment scopes is the same: they are exactly parallel to the polar axis and are sighted on Polaris. The scope either has rotating reticle or is itself rotatable, and has a marker which is aligned with a sidereal time scale (or local mean time and date scales) on a bezel. Polaris is then centred in the appropriate place in the reticle.

For mounts without alignment scopes, but with setting circles, the method is an enhancement of the "rough method":

- Level the base of the mount.
- Adjust the altitude of the polar axis so that its angle to the horizontal is equal to your latitude. Either a wedge (whose angle is that of your latitude) and spirit level, or a plumb-line and protractor, are useful for this if there are no suitable markings on the mount.
- Set up the mount so that the polar axis is pointing as nearly north as you can judge.
- Using the setting circles, set the telescope to the RA and Dec of Polaris.
- Sight through the telescope, adjusting the altitude and azimuth of the polar axis until Polaris is centred in the eyepiece. Ensure that the base of the mount remains level.
- Once Polaris is sighted, offset to $3/4°$ in the direction of Kochab (β UMi). If you know the field of view of your eyepiece, this $3/4°$ should be relatively easy to judge accurately.

If you do this carefully, this method should allow long periods of observing and will be sufficiently precise for piggy-back photography for up to exposures at least as long as 30 minutes with lenses up to 200 mm focal length. It will also allow the setting circles to be used to find objects.

2.1.3 What is the "accurate method"?

If you wish to do prime focus photography through your telescope, or if you are setting up the mount permanently, you should take the time to polar align it as accurately as you possibly can. Firstly, align the mount as accurately as you are able, using the appropriate method above. The instructions given are for the

Northern Hemisphere and assume that you have a high-power guiding eyepiece, giving about $200 \times$ magnification.

Correct any Index Error of the Declination Circle (German Equatorials)

- Use an eyepiece with cross-hairs or a reticle.
- Select a star close to the meridian, i.e. one whose altitude will not change appreciably for a minute or so.
- With the telescope east of the mount, bring the star to the intersection of the cross-hairs, clamp the declination axis and read the declination circle.
- Move the telescope to the west of the mount, bring the star to intersection of the cross-hairs, clamp, and re-read.
- If there is a difference, adjust the index or the circle.
- Repeat until there is no index error.
- Repeat on a star of a different declination (difference $> 30°$).
- If neither the index nor the circle is moveable, halve the difference between the two declination readings for the same star, and use this as a correction (with the appropriate sign) to future declination readings.

Correct the Altitude of the Polar Axis

- By trailing a star on the equator and the meridian, orientate the eyepiece reticle markings/cross-hairs so that they run north-south and east-west.
- Choose a star approximately 6 h east of the meridian within a few degrees of the equator.
- Bring the star to the intersection of the cross-hairs.
- Clamp the Dec axis and track in RA.
- If the star appears to drift northwards, decrease the elevation of the polar axis.
- If the star appears to drift southwards, increase the elevation of the polar axis.
- Make the necessary adjustments and repeat until there is no discernible drift for at least 5 minutes.

Correct the Azimuth of the Polar Axis

- Choose a star near the meridian with a Dec of about 30°.
- Bring the star to the intersection of the cross-hairs.
- Clamp the Dec axis and track in RA.
- If the star appears to drift northwards, the polar axis points west of north.
- If the star appears to drift southwards, the polar axis points east of north.
- Make the necessary adjustments and repeat until there is no discernible drift for at least 5 minutes.

After correcting the azimuth, check the altitude again, and so on until the mount is properly polar aligned.

2.2 Using Setting Circles

2.2.1 What are setting circles?

Setting circles found on the telescopes used by beginning amateurs are usually 360° protractors concentric with the polar (right ascension) and declination axes of an equatorial mount. More advanced telescopes may have digital setting circles, which give a digital read-out of right ascension and declination on an LED or similar display and which work off encoders attached to the axes of the mount. We are here only concerned with the former kind.

2.2.2 How do I set the circles?

The first thing to do is to polar-align the mount. If is properly polar-aligned, there will be no need to set the declination circle (see Section 2.1).

To set the right ascension circle, point the telescope to any part of the sky of known RA. This can either be an object (normally a bright star) of known RA, or the meridian. The RA of the meridian is the Sidereal Time. It is usually more convenient to point it at an object. Slip the RA circle until it reads the RA of the object.

2.2.3 How do I find objects using the circles?

With the circles properly set on an object, simply swing the telescope until the circles read the RA and Dec of the target object. The object should be visible in a low-power eyepiece. With undriven circles you should verify that the RA is correct before moving on to a subsequent target object.

2.2.4 Why does my telescope move the wrong way?

There are two possible reasons for this:

1. Your RA circle is marked for use in both hemi-spheres and you are using the wrong markings. Use the alternative one and continue as above.

2. Your RA circle is designed to be used in the other hemisphere. This is a problem that affects mainly, but not exclusively, Southern Hemisphere observers. Use it as follows:

- Point the telescope to the object of known RA.
- Set the RA circle to the RA of the target object.
- Swing the telescope until the RA circle reads the RA of the original object.
- The declination circle is set and used normally, as described above.

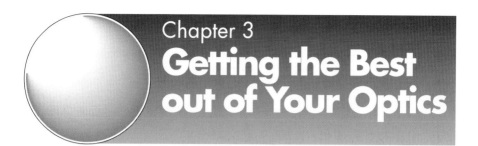

Chapter 3

Getting the Best out of Your Optics

3.1 Telescope Function

To get the best out of a telescope for astronomy, it helps if you understand how it works. A telescope, when used for astronomy, serves three functions: it magnifies, it increases light grasp, and it increases resolution.

3.1.1 How can I determine the magnification?

Magnification is defined as (angle subtended by image)/(angle subtended by object). In an astronomical telescope this is almost exactly the same as (focal length of objective)/(focal length of eyepiece). See Figure 3.1.

3.1.2 What is the lower limit to magnification?

At lower magnifications the exit pupil is larger (see Figure 3.2).

One definition of the least useful magnification is when exit pupil diameter = eye pupil diameter (d). If the exit pupil > d, this is equivalent to the objective being stopped down by pupil. It is, of course, perfectly

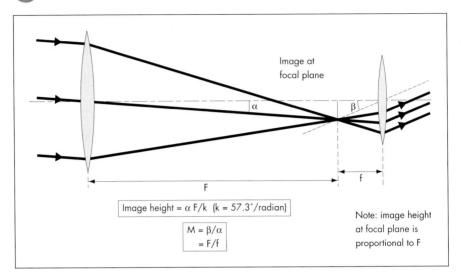

Image at focal plane

α

β

F

f

Image height = α F/k (k = 57.3°/radian)

$M = \beta/\alpha$
$= F/f$

Note: image height at focal plane is proportional to F

Figure 3.1 Magnification

possible to reduce the magnification even further, but extended objects will not get brighter. Also, if magnification is reduced too far in an obstructed telescope (e.g. Newtonian or Cassegrainian), the obstruction may become visible.

Hence lower useful limit of magnification, M_{low}, is given by:

$M_{low} = D/d$ (where D = diameter of objective)

For a middle-aged adult, d ~ 0.2 in; $M_{low} = 5 \times D$, where D is in inches. [d ~ 5 mm; $M_{low} = D/5$, where D is in mm].

3.1.3 What is the upper limit to magnification?

Increasing the magnification reduces the size of the exit pupil (and some of the aberrations of the eye's lens). Decreasing the exit pupil below about 0.03 in (0.75 mm) leads to a progressive impairment of vision.

The theoretical upper limit of magnification, M_{high}, is therefore given by:

$M_{high} = D/0.03$ (D in mm) or $M_{high} = \sim 30 \times D$ (D in inches)

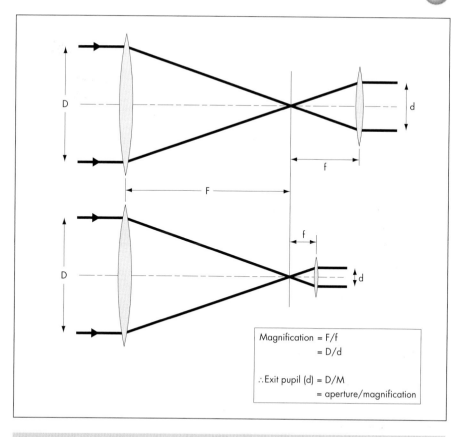

Figure 3.2 Exit Pupil Size

It is also limited by the quality of the objective/ primary and by the optical quality of short focal length eyepieces, which must have highly curved surfaces.

The impairment caused by small exit pupils may not affect vision so adversely as to preclude higher magnifications – indeed, very small exit pupils may attenuate the eye's aberrations. The "rule of thumb" is $50 \times D$ (which is often the most, or more than, the atmosphere will permit), although over $70 \times D$ may be possible sometimes for close double stars. Some observers use up to about $140 \times D$, but these high magnifications probably compensate for a lack of visual acuity. Very high magnifications may be useful when averted vision is used for the same reason (reduced acuity).

3.1.4 What magnification will fully resolve the image?

The minimum magnification, M_r, for at which the image at the focal plane is fully resolved by the eye is $13 \times D$ inches (or the radius of objective in mm). The comments above on increasing magnification to compensate for less-than-perfect visual acuity apply here as well.

3.1.5 What is "light grasp"?

Light grasp is the quantity of light concentrated by objective at primary focus – it is solely a function of D, not of focal length or magnification. It is more meaningfully defined as relative intensity of light from point source passed to retina by optical train, and by naked eye.

If eye's pupil, δ, is not smaller than exit pupil, then light grasp is $D^2/\delta^2 t$; where t is the transmission factor of instrument.

3.1.6 Why are tables of light grasp misleading?

- Observations and assumptions from which the table was constructed may be inapplicable to other conditions.
- Atmospheric seeing can render small instruments able to see fainter objects than larger ones.
- Most telescopes and all observers are not "normal".
- Magnification will affect the outcome.
- Direct or averted vision.
- Type of telescope (loss of light in secondary spectrum of refractors; degradation by obstruction in reflectors).
- Bright field objects affect dark adaptation.

Theoretical limiting magnitude can be calculated. For under magnitude 6.5 skies:

$$\text{theoretical limiting magnitude} = 6.5 - 5\log(\delta) + 5\log(D)$$

(This is low by about 1.5 magnitudes for apertures < 20 in.)

3.1.7 What is the effect of light grasp on extended objects?

No passive optical instrument can increase the brightness per unit area of an extended object.

Light passed to image at prime focus will be D^2t/δ^2 times as great as that received by the retina of the naked eye.

When the telescope is used with magnification, M, the illuminated area of the retina is increased by M^2. Hence the apparent brightness of the image is increased by:

$$(D^2t)/(M^2\delta^2)$$

But the minimum magnification, M', to utilise the light grasp of the objective is D/d.

Substituting for M in the above equation, we find that the telescopic image is t times the brightness of the naked eye image. Since $t < 1$, the brightness of an extended object cannot be increased by a telescope.

The above equation also tells us that increasing the magnification will cause an extended object to dim. The sky is an extended object, hence increasing the magnification of the telescope can increase contrast between point objects (stars) and the sky.

3.1.8 What is resolution?

Diffraction affects the edge of a pencil of light. As a consequence, some light from a point source is spread away from the image. Destructive and constructive interference results in a disc of light (the Airy disc) surrounded by diffraction rings of decreasing brightness (see Figure 3.3).

Assuming the optics are diffraction-limited, the radius of the Airy disc, A, is given by:

$$A = 1.22\lambda/D \text{ radians}$$

where λ is the wavelength of light.

For $\lambda = 500$ nm (yellow light), this reduces to:

$$A \sim 1.22/D \text{ arcsec}$$

Figure 3.3 The Airy Disc

where D is in metres. Thus the only parameter of the telescope which affects the angular size of the Airy disc is the diameter of the objective/primary. Because larger diameter telescopes have smaller Airy discs, not only is the resolution improved, but also the light intensity in the Airy disc is greater, and thus the contrast is improved in larger telescopes.

In practice, the atmosphere usually limits the resolution to anything from 2 to 10 arcsec (or worse).

A typical good site may attain seeing of 1 arcsec.

3.2 Understanding Aberrations

3.2.1 What are aberrations?

Aberrations can be seen as errors in an optical system. There are six optical aberrations, which may affect the image produced by a telescope. Some are uniform across the entire field of view, others affect only abaxial rays of light, i.e. those that are not parallel to the axis of the telescope; some affect the quality of the image, others affect its position. They are:

- chromatic aberration – uniform, error of quality
- spherical aberration – uniform, error of quality
- coma – abaxial, error of quality
- astigmatism – abaxial, error of quality
- field curvature – abaxial, error of position
- distortion – abaxial, error of position

3.2.2 What is chromatic aberration?

Chromatic aberration is an error of refractive systems. Because any light that does not impinge normally on a refractive surface will be dispersed, single converging lenses will bring different wavelengths (colours) of light to different foci, with the red end of the optical spectrum being most distant from the lens. Visible chromatic aberrations can exist in refractor objectives, eyepieces, Barlow lenses and corrector plates.

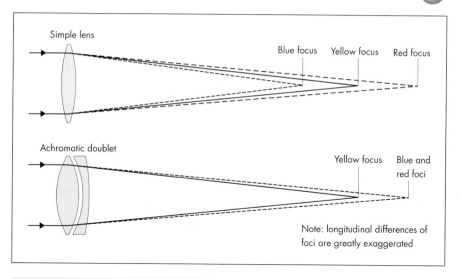

Figure 3.4 Chromatic Aberration

3.2.3 How can chromatic aberration be corrected?

Chromatic aberration can be reduced, but not eliminated, by using multiple lens elements of different refractive indices and dispersive powers. An achromatic lens has two elements and brings two colours to the same focus (see Figure 3.4).

The choice of glass and lens design will determine not only which colours are brought to the same focus, but also the distance over which the secondary spectrum is focused. An apochromatic lens uses three elements and will bring three colours to the same focus.

3.2.4 What is spherical aberration?

Spherical aberration is an error of spherical refractive and reflective surfaces that results in peripheral rays of light being brought to different foci to those near the axis (see Figure 3.5).

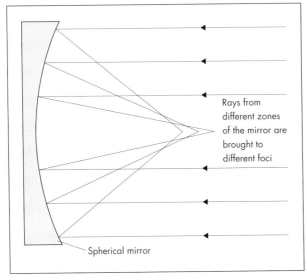

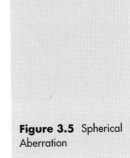

Rays from
different zones
of the mirror are
brought to
different foci

Spherical mirror

Figure 3.5 Spherical Aberration

If the peripheral rays are brought to a closer focus than the near-axial rays, the system is *undercorrected*. If they are brought to a more distant focus, the system is *overcorrected*. Spherical mirrors and converging lenses are undercorrected and diverging lenses are overcorrected.

3.2.5 How can spherical aberration be corrected?

In compound lenses, spherical aberration can be suppressed in the design of the lens by choosing appropriate curvatures for the converging and diverging elements.

In Newtonian mirrors, the spherical aberration is corrected by progressively deepening the central part of the mirror so that all regions focus paraxial rays (rays parallel to the axis) to the same point. The shape of the surface is then a *paraboloid*, that is the surface that results from a parabola being rotated about is axis.

In other telescope systems, the spherical aberration of primary mirrors can be reduced by the introduction of corrector plates or correcting lenses or appropriately curved secondary mirrors or combinations thereof. For example, the ubiquitous Schmidt-Cassegrain and

Maksutov-Cassegrain telescopes use a spherical primary and a corrector plate.

3.2.6 Are there other manifestations of spherical aberration?

Yes. The most common is zonal aberration, in which different zones of the objective lens or primary mirror have different focal lengths.

3.2.7 What is coma?

Perhaps the easiest way to understand coma is to view it as lop-sided spherical aberration (see Figure 3.6). If a primary mirror or objective lens is corrected for paraxial rays, then any abaxial ray cannot be an axis of

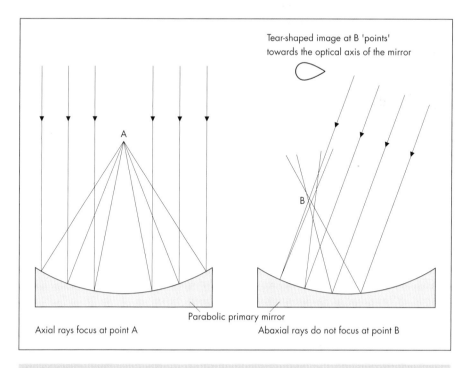

Tear-shaped image at B 'points' towards the optical axis of the mirror

Axial rays focus at point A

Parabolic primary mirror

Abaxial rays do not focus at point B

Figure 3.6 Coma

revolution for the mirror surface and different parts of the incident beam of which that ray is a part will focus at different distances from the mirror or lens. The further off-axis the object, the greater the effect will be. The resulting image of a star tends to flare away from the optical axis of the telescope, having the appearance of a comet, from which the aberration gets its name. It is particularly apparent in short-focus Newtonians.

3.2.8 How can coma be corrected?

In objective lenses, coma can be reduced or eliminated by having the coma of one element counteracted by the coma of another.

A spherical mirror will not exhibit coma since it has no unique axis of revolution. As Bernhard Schmidt demonstrated, a spherical mirror would be ideal if its spherical aberration could be reduced – Schmidt achieved this with a corrector plate. The coma of a paraboloid primary may be removed by using a corrector plate, such as a Ross corrector, which has equal and opposite coma to that of the mirror. Some telescopes also have coma-correcting lens systems.

3.2.9 What is astigmatism?

Astigmatism results from a different focal length for rays in one plane as compared to the focal length of rays in a different plane. A cylindrical lens, for example, will exhibit astigmatism because the curvature of the refracting surface differs for the rays in each plane and the image of a point source will be a line.

Astigmatism will therefore result from any optical element with a surface that is not a figure of revolution.

It can also occur in surfaces that are figures of revolution. Consider two mutually perpendicular diameters across a beam of light impinging obliquely upon a mirror or lens surface. The curvature of the mirror under one diameter differs from that under the other, and thus astigmatism will result. A common source of astigmatism in Newtonian reflectors is slight curvature of the secondary "flat".

Astigmatism is not normally a problem in telescopes used for visual work, but it can become a problem in wide-field astrographs and can compromise the accuracy of astrometric work.

3.2.10 How can astigmatism be corrected?

If the astigmatism results from a faulty optical component, it must either be replaced or have its surface refigured.

Otherwise, astigmatism can be corrected by an additional optical element that introduces equal and opposite astigmatism.

3.2.11 What is field curvature?

No single optical surface will produce a flat image – the image is focused on a surface that is a sphere that is tangential to the focal plane at its intersection with the optical axis.

This is not normally a problem for visual work, although it can be irritating, but will mean that photographs taken with cameras which have flat focal surfaces, or images taken with CCD cameras, cannot be in focus over the entire field.

3.2.12 How can field curvature be corrected?

There are a number of proprietary field flatteners, often combined with focal reducers for photographic work, available from astronomical retailers. Some specialised astrographs, such as the Wright camera, have optical components that give focal surfaces of such very long radius that they can sensibly be considered to be flat. Others use film holders that are themselves curved.

3.2.13 What is distortion?

Distortion is an aberration by which a square object gives an image with either convex lines, called *negative*

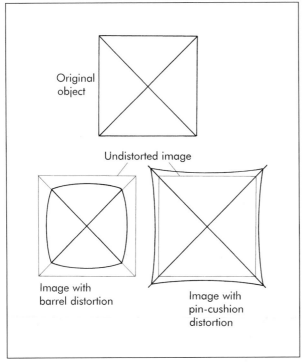

Figure 3.7 Distortion

or *barrel distortion*, or concave lines, known as *positive* or *pincushion distortion* (see Figure 3.7).

It results from differential magnification at different distances from the optical axis. In visual telescopes, distortion from the objective lens or primary mirror is negligible, but it can occur in eyepieces.

3.2.14 How can distortion be corrected?

Because any distortion almost certainly comes from the eyepiece, the simplest way to overcome it is by careful eyepiece selection.

3.3 Basic Star Testing

3.3.1 What is a star test?

A star test uses focused and defocused images of a bright star in order to obtain information about the quality of a telescope's optics. It is solely a qualitative, not a quantitative, test, although experienced testers are able to make quantitative inferences from it. However, your primary concern is the quality of the optics, not the numerical wave-ratings with which the vendor advertised his wares. Star testing is very sensitive and telescopes that show slight aberrations under a star test can still produce images that are indistinguishable from those obtainable with supposedly perfect optics.

3.3.2 How do I perform a star test?

Firstly, you need to wait until the atmosphere is reasonably still – if conditions are poor, it is more than a little difficult to interpret what you see (more on recognising bad viewing conditions later, although the obvious first check is to see if the stars are twinkling/scintillating with naked eye, then check with binoculars). Let your telescope reach thermal equilibrium with the outside air. You need to remember that the entire optical system between the star and your brain includes the atmosphere and your eye, as well as the telescope.

Use an eyepiece that gives a magnification in the region of ×50 per inch of aperture, with a maximum of around ×200, and focus the chosen star. If you don't have a driven telescope, use Polaris if you live in the Northern Hemisphere, otherwise choose a bright star and centre it in the field of view. One of the difficulties people often have with star tests is that they choose an insufficiently bright star.

Can you get it to snap to a clearly defined focus? If not, either conditions are poor or there is something wrong with the telescope (could be the optics or collimation thereof). You may be able to see rings around the focused image – this is not unusual.

Rack/turn the focuser a little inside and outside focus. The star image appears to become a larger disk which, on closer examination, is actually a dark "hole" surrounded by rings. How close an examination depends on the optical system – with good skies, optics, and eyes the rings are immediately very obvious. In a refractor there will not be the dark hole, which is an effect of the obstruction to the light path caused by a reflector's secondary mirror. In a perfect telescope, the rings will be identical inside and outside focus.

Once you have learned to see these rings, repeatedly move quickly from one to the other (i.e. inside to outside focus) and back again – this has a similar effect to a blink comparator and will more easily enable you to see differences.

3.3.3 How do I star-test a refractor?

A refractor is star-tested in essentially the same way as a reflector but, because of the chromatic aberration which can make the results difficult to interpret, especially on intrafocal images, you should use a yellow or green filter (Wratten #15 or #58), which will filter out the secondary spectrum.

3.3.4 What will I see with perfect optics?

A perfect system shows a series of rings of equal brightness (see Figure 3.8a).

3.3.5 What do tube currents look like?

If you see a wobble in the extrafocal image, or if it has an irregular sort of blobby turbulent appearance, it is probably tube currents in the telescope – you need to wait until it settles down.

3.3.6 What can I do about tube currents?

Wait until the telescope reaches thermal equilibrium with the air.

3.3.7 What does atmospheric turbulence look like?

If it looks a bit like you are watching the riverbed through a flowing river, it is atmospheric turbulence, i.e. bad seeing. The difference between what you see with tube currents and with atmospheric turbulence is that the "ripples" from the atmosphere tend to move in the same direction. You will probably need to wait until the atmosphere is less turbulent.

3.3.8 What does coma look like?

Coma will exist in any non-spherical (e.g. paraboloidal) primary mirror (see Figure 3.8b). Unless your scope is properly collimated you may see coma at the centre of the field; i.e. star images which appear as mini comets. They "point" towards the optical axis of the instrument. The intra- and extra-focal images show the rings as brighter on one side than the other. The obvious thing is to collimate properly, so that a star in the centre of the field shows no coma.

3.3.9 What can I do about coma?

The obvious thing is to collimate properly, so that a star in the centre of the field shows no coma (see Section 3.4).

3.3.10 What does astigmatism look like?

Astigmatism shows as elliptical rings – the orientation of the major axis of the ellipse moves through 90° between intra- and extra-focal images (see Figure 3.8c). Their actual orientation in the eyepiece may vary from that shown in the drawing.

3.3.11 What can I do about astigmatism?

Astigmatism can result from optical defects or from poor collimation, although in the latter case it would probably be disguised by the coma that would also result. Rotate each optical element (lens or mirror)

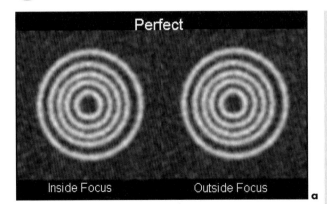

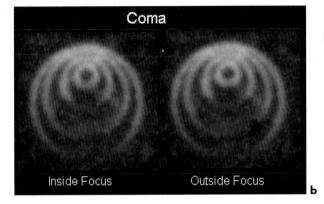

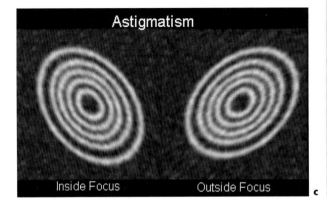

Figure 3.8
Diffraction Rings as seen
in a Star Test.
a Perfect optics
b Coma
c Astigmatism

3

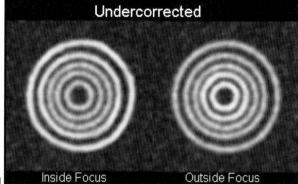

d Undercorrected

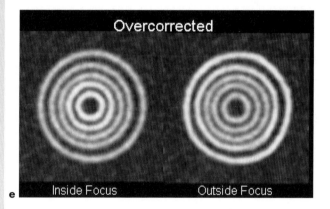

e Overcorrected

d Undercorrected
e Overcorrected
f Turned Edge

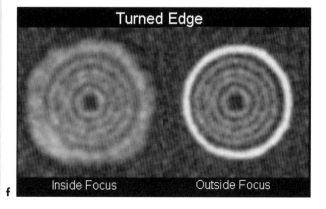

f Turned Edge

separately. If the axes of astigmatism change, there is astigmatism in the rotated element. To reduce the astigmatism that element must be corrected (re-assembled correctly, refigured, or replaced).

3.3.12 What does under-correction look like?

An undercorrected mirror has a bright outer ring inside focus and a bright inner ring outside focus (see Figure 3.8d).

3.3.13 What can I do about undercorrection?

The offending optical element must be corrected (refigured) or replaced.

3.3.14 What does over-correction look like?

An overcorrected mirror shows the reverse of the above – there is a gradation of brightness through the rings (see Figure 3.8e).

3.3.15 What can I do about overcorrection?

The offending optical element must be corrected (refigured) or replaced.

3.3.16 What does a "turned edge" look like?

If you have an extrafocal image which has a bright outer ring (as does an overcorrected mirror) which is significantly brighter than the next one (i.e. not a gradation of brightness through the rings), you probably have a turned edge. The intrafocal image will not show a change in brightness and the outer ring appears fuzzy (see Figure 3.8f).

3.3.17 What can I do about a turned edge?

The simplest option is to mask the edge of the mirror (turned edge is almost exclusively associated with primary mirrors) with an annular mask. If you do not wish to mask it, then it must be corrected (refigured) or replaced.

3.3.18 Where can I get more information?

H.R. Suiter's excellent *Star Testing Astronomical Telescopes* is the acknowledged "bible" on the subject.

3.4 Collimation

3.4.1 What is collimation?

Collimation is the adjustment of the optical elements of a telescope so that they are in correct optical alignment. Specifically, the optical axis of the eyepiece must be colinear with the optical axis of the primary mirror or objective lens at their mutual focus.

Telescopes which are moved around a lot, i.e. most amateur instruments, tend to require recollimation more frequently than do permanently mounted telescopes, owing to the jarring that occurs when the telescope is transported. In general, refractors hold their collimation better than reflectors and catadioptrics.

Poorly collimated instruments suffer from poor image quality, so it is important to learn how to do it properly.

3.4.2 How can I tell if my telescope needs collimation?

If there are comatic images at the centre of the field of view, collimation is certainly required. A star test will reveal even the slightest mis-collimation – the intra- and extra-focal diffraction rings will not be concentric circles, owing to coma, or astigmatism, or both.

3.4.3 How do I collimate a Newtonian telescope?

The optics of a properly collimated Newtonian telescope meet three requirements:

- The optical axis of the eyepiece is colinear with the optical axis of the primary mirror at their mutual focus.
- The optical axis is coincident with the optical centre of the secondary mirror.
- The optical axis is reflected at exactly 90° by the secondary mirror.

In addition, it is wise to adhere to an optical-mechanical condition, i.e. that the optical axis of the primary mirror is colinear with the mechanical axis of the tube. If this condition is not met, not only may the aperture cause some vignetting, but the optical axis will not describe a great circle in the sky when the telescope is moved in declination or altitude and, if the tube is rotated in its cradle or rings, the direction of the optical axis will change.

Collimation should be done in seven steps. The first five are more easily done in daylight:

1. Square the focuser to the tube. Mark a spot on the tube diametrically opposite the centre of the focuser and use a sighting tube to align the focuser with it. You can make simple sighting tube for a 1.25 in focuser with a 35 mm film can with 1 mm holes in the centre of the base and the cap.

2. Correctly position the secondary mirror. First, centre the secondary mirror in the tube by adjusting the spider or stalk. Note that it is the optical centre, not the physical centre, which should be centred. If necessary, make a tiny mark at the optical centre of the secondary (see Figure 3.9).

The offset of the physical centre away from the focuser (transverse offset) can be calculated by the formula:

$$\text{Offset} = \text{minor axis} \div (4 \times \text{focal ratio})$$

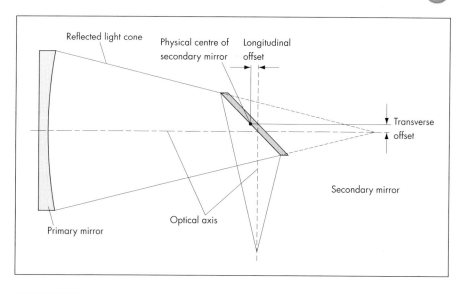

Figure 3.9 Diagonal Offset

Hence for a 36 mm minor axis secondary used with a f/8 telescope, the offset is

$36 \div (4 \times 8)$ mm $= 36 \div 32$ mm $= 1.125$ mm.

The same secondary with an f/4 telescope will have an offset of 2.25 mm – from this it is clear that, whilst offset can be ignored at large focal ratios, it is important at small focal ratios.

3. Once the secondary is centred in the tube, adjust it by ensuring that the reflection of the primary mirror is central in the secondary. This will only occur when the secondary is correctly offset along the tube (longitudinal offset), so this longitudinal adjustment occurs automatically.

4. Adjust the angle of the secondary mirror until it appears circular. If you have to make a large adjustment at this stage, repeat steps 2 and 3.

5. Adjust the primary mirror until the image of the focuser drawtube is in the centre of the secondary mirror (see Figure 3.10).

6. Check that the optical axis is approximately colinear with the axis of the tube. If no part of the edge of the aperture is visible from the drawtube, then no vignetting by the telescope tube will occur.

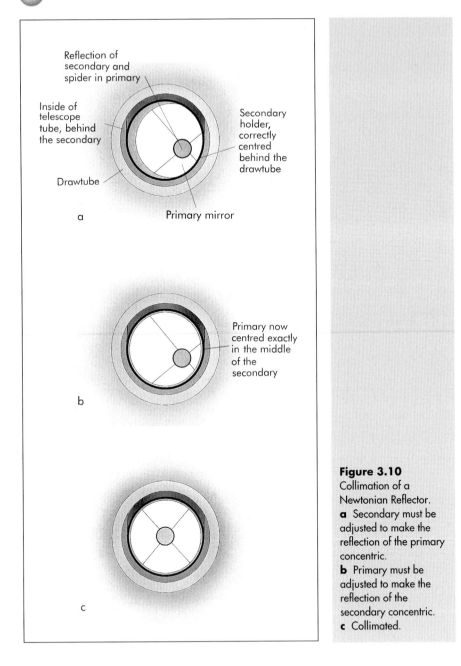

Reflection of
secondary and
spider in primary

Inside of
telescope
tube, behind
the secondary

Secondary
holder,
correctly
centred
behind the
drawtube

Drawtube

a

Primary mirror

Primary now
centred exactly
in the middle
of the
secondary

b

c

Figure 3.10
Collimation of a
Newtonian Reflector.
a Secondary must be
adjusted to make the
reflection of the primary
concentric.
b Primary must be
adjusted to make the
reflection of the
secondary concentric.
c Collimated.

7. Star test the telescope. The purpose of collimation is
   to give the best possible images of stars and other
   celestial objects. Make sure you test on a star in the
   middle of the field of view (see Section 3.3).

3.4.4 How do I collimate a refractor?

Many amateur refractors are assumed by the manufacturer to be permanently collimated when they are shipped to the distributor, and thus they may not have facilities for recollimation. Therefore, some or all of the following may be inapplicable to your telescope and you would be wise to establish just what is possible before you begin. You should also check with the vendor and/or manufacturer that any actions you propose to take would neither invalidate any warranty nor be irreversible.

There are only four steps in the collimation of a refractor:

1. Centre and square the focuser. This is not usually possible with commercial amateur instruments, and should anyway be unnecessary in a well-made instrument. If it is both necessary and possible, the simplest way is to remove the objective lens with its cell and place a pair of cross-hairs over the aperture of the telescope tube so that they cross in the centre. Centre the focuser and use a peephole sight tube to align it on the cross-hairs. You can make simple sighting tube for a 1.25 in focuser with a 35 mm film can with 1 mm holes in the centre of the base and the cap.

2. Centre the objective lens. If this is possible, it is usually achieved by rotating an eccentric ring in the cell. This may require a special tool (e.g. a peg spanner) and you may have to slacken grub-screws before the ring will rotate. You can mark the centre of the lens with a small disc, such as is produced by a hole-punch, of acid-free tissue which is dampened with clean water to enable it to be stuck to the glass. The tissue disc should be aligned with the peephole sight tube.

3. Square on the objective lens. There are often either push-pull screw pairs or single screws with springs for this purpose. Fix a small light source (e.g. penlight or LED) such that it is a few centimetres outside the focuser and exactly along its axis. The light will be reflected off the surfaces of the lens elements. Introduce a thin piece of glass, such as a microscope slide cover slip, at 45° between the light

and the focuser, so that you can see the reflections off the objective. When the objective is squared, the reflections will be coincident.

4. Star test the telescope.

3.4.5 How do I collimate a catadioptric?

It is a misconception that Schmidt-Cassegrains and Maksutov-Cassegrains do not require collimation. This is incorrect, especially as these telescopes are extremely sensitive to miscollimation.

The only possible adjustment is the three screws (or three pairs of screws) which adjust the angle of the secondary mirror. (In Maksutov-Cassegrains whose secondary is an aluminised spot on the corrector plate, this adjustment is not possible.) The process of collimation is one of "trial and improvement", star-testing after each adjustment. The process is essentially simple, but some precautions do need to be taken:

- Do not adjust the central screw – this secures the secondary mirror.
- Adjust the screws in small increments – half a turn at most – being very careful not to over- or under-tighten them.
- When one screw is loosened, the other two must be tightened, and vice-versa.
- Re-centre the star in the field of view after each adjustment.

3.5 Cleaning the Optics

3.5.1 What is the "golden rule" of cleaning optics?

The golden rule is "prevention". Always keep dust caps on apertures and focuser drawtubes (a 35 mm film-can makes a perfect dust-plug on 1.25 in drawtubes) when the instrument is not in use. Do not touch optical surfaces except when necessary; in particular, don't get eyelash grease on your eyepieces.

3.5.2 What is the "silver rule" of cleaning optics?

The silver rule is "don't"! The damage you may do to a surface in an attempt to clean it could be more deleterious to the image than the dirt being removed.

3.5.3 What is the "bronze rule" of cleaning optics?

The bronze rule is that you should always consult the manufacturer or supplier before attempting to dismantle your telescope and before applying any substances to the optical surfaces.

3.5.4 What is the best way to remove dust?

Use a camera puffer-brush, but don't actually brush the dust unless it cannot just be blown off. Blow the dust to the edge of the surface then, if necessary, flick it off with the brush. Don't use canned air unless it is designed for the purpose – the propellant can be deleterious to the mirror or lens coatings.

3.5.5 How often should I clean my optics?

As infrequently as possible. A mirror or lens can be truly filthy yet the telescope can still give good images. As with so many things, prevention is better than cure. If a telescope is properly looked after, its optics can go for many years without being cleaned.

3.5.6 What if my mirror is really dirty?

For all mirror-cleaning operations, ensure that the mirror can only fall onto a soft surface, should you drop it. Clean folded towels or blankets are suitable for this purpose.

If the mirror is truly and obscenely filthy, and you really *have* to clean it, carefully take it out of its cell and rinse it in running water. If that doesn't remove sufficient muck, fill a basin with water and a little bit of real soap powder/flakes (i.e. not detergent). Put the mirror in and swill it around. With the mirror surface still under water, gently (i.e. *absolutely* no pressure) swab the surface of the mirror from centre to edge with lint-free cotton swabs, rolling the swab in the direction of travel, in order to rotate any grime away from the optical surface. Do not use synthetic cotton balls – they may scratch the surface. You would be advised just to do a little bit in the centre first and examine the swabbed area with a magnifier or eyepiece to make sure that you are not causing any damage. Rinse the mirror under copious amounts of running tap water.

Then rinse the mirror with distilled water, particularly if you live in a hard water area. Water sold for topping up lead-acid batteries is good for this purpose. Then leave it on its edge to drain and dry. At all times take care not to touch the surface, and make sure it is safely supported and protected whilst drying.

3.5.7 What if my refractor objective is really dirty?

There is a substance on the market called "collodion", which is excellent for use on refractive surfaces, but do check with the manufacturer before using it on an expensive optic! Collodion is sprayed onto the surface, where it forms a skin. The skin is peeled off, taking with it the grime from the surface.

In the absence of such a product, you can make a cleaning solution as follows:

6 parts distilled water (that sold for battery top-up will suffice)
1 part pure isopropyl alcohol
2 drops liquid detergent (e.g. mild washing-up liquid).

First, remove all dust (see Question 3.5.4). Dampen a lint-free pure cotton swab in the solution and gently (i.e. weight of swab alone) swab from centre to edge, rolling the swab in the direction of travel, in order to rotate any grime away from the optical surface. Dry the lens by gently blotting it with a dry cotton swab or a piece of lens tissue.

3.5.8 How do I clean a dirty Schmidt or Maksutov corrector?

Treat these in the same way as an object glass. Be particularly careful not to scratch the surface with grit from the edge of the secondary holder.

3.5.9 How do I clean dirty eyepieces?

Treat these in the same way as an object glass.

3.5.10 How do I clean dirty binocular lenses?

Treat these in the same way as an object glass.

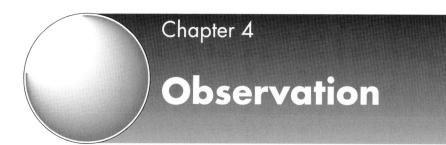

Chapter 4

Observation

4.1 Observing Criteria

4.1.1 What constitutes good observing conditions?

The three components of good observing conditions are transparency, lack of sky brightness, and seeing. Understanding these components and the effect they have on observing will help you get the most out of your telescope.

4.1.2 What is transparency?

Transparency is the optical clarity of the atmosphere and is affected by clouds, moisture, and other airborne particulates, such as industrial pollutants and volcanic matter. Hazy skies and thin cirrus cloud absorb more light than clear skies, making fainter objects less visible and reducing the contrast of brighter objects.

The sky is often at its most transparent after rain (which washes out particulates), especially when the atmospheric pressure is high.

4.1.3 What is sky brightness?

Ideal conditions for observing are when are when the night sky is velvet-black. Sky brightness can be caused by the Moon, aurorae and light pollution. Although it is often not a hindrance to the observation of bright objects, such as the Moon or the naked-eye planets, bright skies reduce the contrast of fainter extended objects, especially nebulae, making them difficult, if not impossible, to observe. Light Pollution Reduction (LPR) filters improve deep sky viewing from light-polluted sites by attenuating some unwanted light and selectively transmitting light from certain deep sky objects.

4.1.4 What is seeing?

Seeing refers to atmospheric and directly affects the quality of detail visible in extended objects. The air in our atmosphere refracts incoming light, the amount of refraction being dependent on air density layers of different temperatures have different densities and, in a turbulent atmosphere, can cause stars to scintillate or become "blobby" and the edges (limbs) of the Moon and planets to appear to "boil". The diameter of the "cell" of air through which you observe is that of the aperture of your instrument, which is why larger telescopes appear to be more sensitive to poor seeing than do those of lesser aperture – there is more chance of the air in a smaller cell having longer periods of stability. Under good seeing conditions and with a good instrument, exquisite detail is visible on the outer bright planets, especially Jupiter and Mars, and stars are small, stable pinpoint images.

The best seeing is often to be had during the still, misty nights of spring and autumn (fall).

4.1.5 What conditions do I need to observe the Moon?

The best time to see details on the lunar surface is when they are sunlit in profile, i.e. during the partial phases. Low powers will reveal all or most of the lunar disk within the telescope's field of view. Higher powers will reveal more detail, but will be limited by seeing conditions – when the atmosphere is turbulent, the lunar limb will appear to "boil".

Yellow filters work well at improving contrast. Neutral density and polarizing filters reduce glare.

4.1.6 What conditions do I need to observe the planets?

Atmospheric conditions are usually the limiting factor on the visibility of planetary detail. For this reason they should be observed when they are as high as possible in the sky and the optical path between you and the planet should not be over heat sources, such as domestic rooftops. Because they are generally bright objects, transparency is less important than good seeing.

The planetary discs of the superior planets are greatest when they are near opposition so, on balance, more detail will be possible at this time.

4.1.7 What conditions do I need to observe the Sun?

In order to see as much detail as possible on the Sun's surface, good seeing is needed. This is most likely to be in early morning and late afternoon, when the air temperature is close to that of the ground, making turbulence less likely.

4.1.8 What precautions should be taken for solar observation?

The Sun is the only object that it can be dangerous to observe. To observe it safely, you must either project its image onto a screen or use a full-aperture filter designed specifically for the purpose of solar observations. On no account should you use either an eyepiece filter or a Herschel wedge.

If you have a catadioptric telescope, do not use it for solar projection – the build-up of heat inside the telescope could be damaging or dangerous.

Always ensure that the finder of the telescope is capped during solar observation.

To align the telescope on the Sun, observe the telescope's shadow and adjust the telescope until the shadow is a perfect circle.

4.1.9 What conditions do I need to observe Deep Sky Objects?

Deep Sky Objects (DSOs) are those outside the solar system. They include star clusters (globular and open), planetary nebulae, diffuse nebulae, double stars, supernova remnants, and. They often have a large angular size and low surface brightness. Therefore large aperture and low- to medium-power is suitable for their observation. They should be observed from a dark site.

Light pollution makes many deep sky objects difficult, if not impossible, to observe. Light Pollution Reduction (LPR) filters help reduce the background sky brightness, thus increasing contrast and rendering more objects visible (see Question 1.4.1).

Oxygen III (O-III) filters are ideal for planetary nebulae (see Question 1.4.1).

4.1.10 What tricks can I use to see fainter objects?

- Averted vision. The human eye is most sensitive to light at the periphery of the field of vision so, if you look to the side of the object so that it is near the edge of your field of view, it may pop into sight. Using averted vision on objects you can see without it may enable you to see detail that was otherwise invisible. Using my right eye, I avert my eye slightly down and to the left to get the best results, but experiment to find which direction works best for you.
- Darkness. Make sure your eyes are properly dark-adapted. Full dark adaptation involves the production of a chemical, visual purple, in the retina, and takes at least twenty minutes to attain. Light can destroy it in a fraction of a second. Some observers find that wearing an eye-patch before an observing

session is helpful. Even faint stray light, insufficient to destroy your dark adaptation, can be a hindrance to observation, so you must try to eliminate it. Eyecups on the eyepiece are very helpful. An observing hood that covers your head and the eyepiece makes a tremendous difference.

- Tap the telescope. Your eye is sensitive to movement, particularly at the periphery of vision. Tapping the telescope whilst using averted vision can sometimes cause an object to pop into view. However, a constantly moving telescope (such as is mounted on an inadequate mount) is a hindrance, not an aid, to observation!

- Oxygenate your retina. The retina seems to be more sensitive when its blood supply is more highly oxygenated. Breathe deeply (but make sure you do not hyperventilate to the extent that you pass out!) whilst observing and, if you smoke, abstain during (and for an hour or so before) observing.

- Change eyepieces. Not only are different objects easiest at different magnifications, but the characteristics of different eyepieces mean that the best one for the object may not be the one you are using. At all times make sure you have achieved best focus.

- Patience, persistence and experience are possibly the most effective aids to the observation of very faint objects.

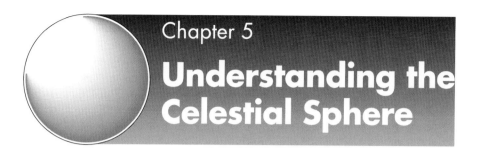

Chapter 5

Understanding the Celestial Sphere

5.1 Positional Astronomy

5.1.1 What is the Celestial Sphere?

We imagine that astronomical objects are projected onto a celestial sphere, centred upon the Earth. The Earth's North and South poles are projected to the North Celestial Pole (NCP) and South Celestial Pole (SCP), respectively. The Earth's equatorial plane passes through the Celestial Equator (see Figure 5.1).

5.1.2 How do we refer to positions on the Celestial Sphere?

Positions on the surface of any sphere are referred to a reference plane and a reference direction. The most common example of this is the Latitude and Longitude system we use to define positions on the Earth's surface. Latitude is referred to a reference plane, an imaginary plane that passes through the Earth at the

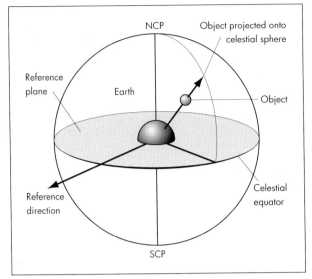

Figure 5.1 The
Celestial Sphere (1)

Equator. Longitude is referred to a reference direction, the meridian that passes through Greenwich.

Similarly, positions on the Celestial Sphere are referred to a reference plane and a reference direction. There are two systems commonly used by observational astronomers: the Horizon Co-ordinate system and the Equatorial Co-ordinate system.

5.1.3 What is the Horizon Co-ordinate system?

An observer on Earth sees part of the curve of Earth's surface as the horizon (see Figure 5.2).

This horizon can also be projected onto the Celestial Sphere as the Celestial Horizon. The point directly above the observer is the Zenith; the point directly below is the Nadir. The observer's compass directions, N, E, S and W, can also be projected onto this plane. A great circle on the Celestial Sphere that passes through N, S and the Zenith is called the Meridian. The meridian also passes through the Celestial Poles (see Equatorial Co-ordinates, Question 5.1.5).

The two elements of the Horizon Co-ordinate System are Altitude and Azimuth. The Altitude of an object is measured with respect to the Celestial Horizon; i.e. its reference plane is the horizontal plane.

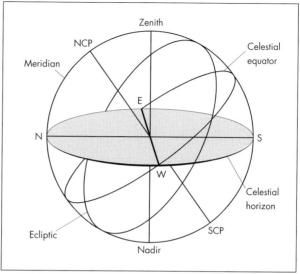

Figure 5.2 The Celestial Sphere (2)

Azimuth is measured through east from North; i.e. its reference direction is the North point on the horizon.

Because the horizon co-ordinates of any astronomical object relate to the position in space of the observer, they are unique to each observer and vary with time.

5.1.4 Where are the Celestial Poles and the Celestial Equator in relation to the Horizon?

The altitude of the NCP is equal to the latitude of the observer. It is due north. The highest altitude of the celestial equator (i.e. due south) is the complement of the observer's latitude (see Figure 5.3). Southern Hemisphere observers should substitute "north" for "south" and vice versa.

5.1.5 What is the Equatorial Co-ordinate system?

Equatorial Co-ordinates are referred to the Celestial Equator as the reference plane and to the First Point of

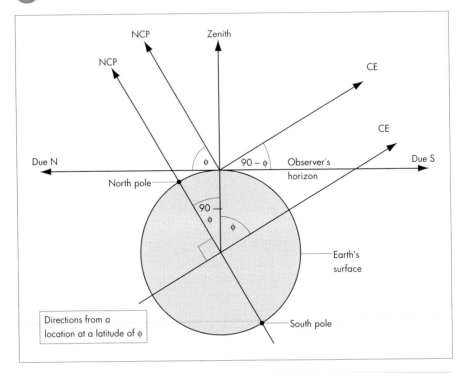

Figure 5.3 Horizon Positions

Aries (FPA) as the reference direction (see Figure 5.4). The FPA is the position of the Vernal Equinox; it is where the Ecliptic crosses the Celestial Equator. Owing to the Precession of the Equinoxes, the FPA moves around the equator with a period of about 26 000 years. Equatorial positions must therefore be referred to a date or Epoch. The epoch currently used is AD2000.0

The declination (Dec) of an object is its position measured northwards from the equatorial plane. It is measured in degrees. Objects south of the equator have negative declinations. Declination is analogous to Latitude on Earth.

The right ascension (RA) of an object is its position, measured eastwards around the equator, from the FPA. By convention, the equator is not divided into 360°, but into 24 hours, with subdivisions of minutes and seconds. (The reason for this will become apparent in the section on Sidereal Time.) Right ascension is analogous to Longitude on Earth.

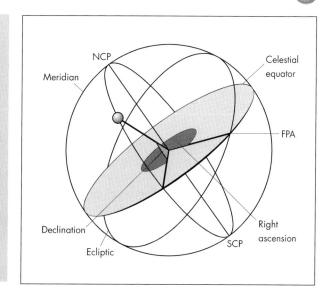

Figure 5.4 Right Ascension and Declination

The Hour Angle (HA) of an object is its position, measured around the celestial equator, westward from the observer's meridian.

5.1.6 How can we convert from one co-ordinate system to another?

Users of altazimuthally mounted instruments often find it useful to be able to convert equatorial co-ordinates to horizon (altitude and azimuth) ones.

sin(a) = sin(d) sin(l) + cos(d) cos(l) cos(LST – r)
cos(A) = (sin(d) – sin(l) sin(a))/(cos(l) cos(a))

where r = right ascension; LST = local sidereal time; d = declination; A = azimuth; a = altitude; and l = latitude.

5.1.7 Where can I find out more?

A number of practical astronomy books cover positional astronomy. Among the best is Chris Kitchin's excellent *Telescopes and Techniques*, which is thorough and explains it in clear language.

Gerald North's chapter in *The Modern Amateur Astronomer* gives a simple introduction to co-ordinate transformations. For more detailed accounts, see Jean Meeus' *Astronomical Algorithms* or Peter Duffett-Smith's *Practical Astronomy with your Calculator*.

5.2 Time

5.2.1 What are Solar and Sidereal Time?

Solar Time is time measured with respect to the Sun. Sidereal Time (from Latin *sidus* = star) is time measured with respect to the celestial sphere.

Viewed from Earth, the celestial sphere rotates through 24 h of RA in a Sidereal Day (see Figure 5.5).

Figure 5.5 Solar and Sidereal Time

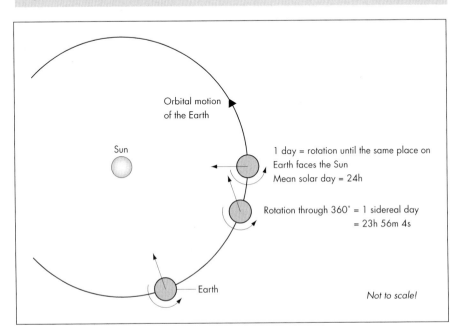

5.2.2 What is Local Sidereal Time?

The Local Sidereal Time (LST) = RA + HA. Greenwich Sidereal Time (GST) is the LST at the Greenwich Meridian.

The Hour Angle (HA) of an object is its position, measured around the celestial equator, westward from the observer's meridian (see Figure 5.6).

5.2.3 How is Mean Solar Time determined?

Owing to the ellipticity of Earth's orbit (see Kepler's Second Law) and the obliquity of the ecliptic, the Sun does not appear to travel through the sky at a uniform rate. Mean Solar Time is referred to an imaginary body, the Mean Sun, which travels around the celestial equator (not the ecliptic!) at a constant rate of 360° per year.

Mean Solar Time is defined as the Hour Angle of the Mean Sun (HAMS) + 12 h.

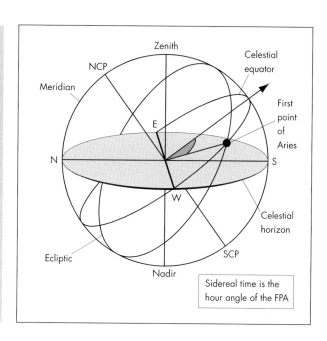

Figure 5.6 Sidereal Time. The Hour Angle of the FPA is shaded.

5.2.4 How is Solar Time determined?

Solar Time is the time given by a sundial, i.e. the time given by the real Sun.

5.2.5 What is the Equation of Time?

Solar Time differs from Mean Solar Time by the Equation of Time (E). E can be as great as 16 minutes. (See Figure 5.7.)

$$E = \text{Solar Time} - \text{Mean Solar Time}$$

The observable effect of this is that the position of the midday Sun at 12:00 clock time (or any other clock time) will describe an irregular lemniscate, called the analemma, throughout the year.

Figure 5.7 Equation of Time

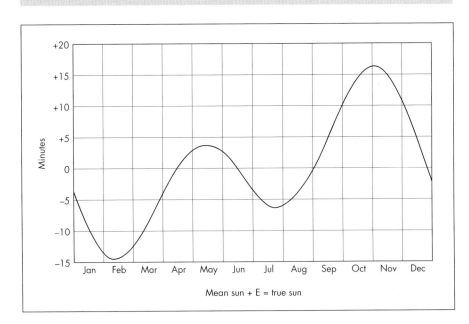

Mean sun + E = true sun

5.2.6 What is Universal Time?

Astronomers use Universal Time (UT). For most every-day purposes, this can be taken as the Mean Solar Time on the Greenwich Meridian (Greenwich Mean Time or GMT). UT is actually calculated from sidereal time. UT0 is the result of this calculation. UT1 is corrected for Earth's "polar wobble" (nutation, precession, etc.). When "UT" is used by astronomers, it normally refers to UT1.

Co-ordinated Universal Time (UTC) is used for time-signal broadcasts. It is derived from International Atomic Time (TAI), from which it differs from a whole number of seconds. It is linked to UT1, from which it never differs by more than 0.9 s – hence the use of leap seconds to keep them in step.

5.2.7 Where can I find out more?

A comprehensive treatment more suitable for the beginning astronomer is given in both Chris Kitchin's *Telescopes and Techniques* and Gerald North's chapter in *The Modern Amateur Astronomer.*

5.3 Interpolating Ephemerides

5.3.1 How do I interpolate an ephemeris?

Interpolation is the calculation of values that are intermediate to the given values in the ephemeris.

Let us consider the declination of Ceres as given in the Feb 1999 ephemeris fragment for Ceres (positions are at 0000 UT). We wish to find the declination at 2100 UT on February 13.

Date	Right ascension	Declination
12 Feb 1999	3 h 57 m 24.84 s	20.97969°
13 Feb 1999	3 h 58 m 03.81 s	21.05635°
14 Feb 1999	3 h 58 m 44.15 s	21.13342°
15 Feb 1999	3 h 59 m 25.84 s	21.21088°
16 Feb 1999	4 h 00 m 08.87 s	21.28870°

Let us label the dates "x" and the declination "y". The value of x corresponding to our required date and time is February 13.875.

We choose the 3 nearest dates and their corresponding values of declination (converted to decimal form):

x	y
Feb 13 (x_1)	21.05635 (y_1)
Feb 14 (x_2)	21.13342 (y_2)
Feb 15 (x_3)	21.21088 (y_3)

$$a = y_2 - y_1 = + 0.07707$$

$$b = y_3 - y_2 = + 0.07746$$

$$c = b - a = + 0.00039$$

$$n = x - x_2 = - 0.125$$

$$y = y_2 + \frac{n(a + b + nc)}{2}$$

$$= 21.13342 - 0.125 (0.07707 + 0.07746 + (- 0.125 \times 0.00039))/2$$

$= \underline{21.12376}$ (This is within half an arcsecond of value computed from orbital elements.)

We can interpolate from three values when the third difference (c) is close to zero. If the third difference cannot be ignored, we must use five values:

x_1	y_1				
		$a = y_2 - y_1$			
x_2	y_2		$e = b - a$		
		$b = y_3 - y_2$		$h = f - e$	
x_3	y_3		$f = c - b$		$k = j - h$
		$c = y_4 - y_3$		$j = g - f$	
x_4	y_4		$g = d - c$		
		$d = y_5 - y_4$			
x_5	y_5				

The interpolation formula is then:

$$y = y_3 + \frac{n(b+c)}{2} + \frac{n^2 f}{2} + \frac{n(n^2 - 1)(h + j)}{12} + \frac{n^2 (n^2 - 1)k}{24}$$

Obviously, a degree of common sense must be used when deciding whether to use 3 or 5 values.

5.3.2 Where can I find out more about interpolation?

Jean Meeus' *Astronomical Algorithms* includes a comprehensive coverage of this.

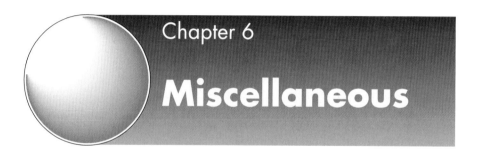

Chapter 6

Miscellaneous

6.1 Counteracting Dew

6.1.1 Why does dew form?

In order to know how to combat dew, it is important to have some understanding of why it forms. Warm air can hold more moisture than cold air. Hence when warmer, moist air cools down, there will come a point where it is saturated and will begin to give up its moisture. The principles of dew reduction are then simple: reduce the amount of cooling of the optical surfaces and reduce the amount of warm moist air (e.g. breath) that comes into contact with them.

6.1.2 What is radiative cooling?

Outside our biosphere is space; far enough out and it is space at a temperature of 2.7 K. Although the effective temperature of the sky is perhaps 100 K or so warmer than that, it is still a great deal colder than the surface of the Earth. Hence, on clear nights (i.e. those good for astronomy) there will be a net loss of heat by radiation from the surface of the earth and things on it, like telescopes. As they cool, they become prone to dew (and frost) formation.

6.1.3　What are the principles of dew prevention?

An observatory will shield the telescope from a great deal of the sky, thereby reducing this radiative cooling. For those of us without observatories, our simplest solution is to reduce the amount of sky that the optical components can "see". Newtonian primaries, particularly those in solid tubes (as opposed to trusses), are usually well shielded. Refractor (including binocular) objectives, Schmidt corrector plates and eyepieces are among the most dew-prone, with *Telrad*® finders heading the list by a long way.

6.1.4　How can I keep dew off objectives and corrector plates?

Dew caps provide the simplest way of shielding object glasses and corrector plates from the cold sky, but very few telescopes are provided with sufficiently long ones. Foam sleeping mats, available from camping suppliers, are easy to cut to size to make a dew cap. The butted ends can be stuck with duct- or gaffer-tape, and the inside can be blackened (*Krylon*® *Ultra Matt Black* spray is good for this). As long as the end of the dew cap is not visible in the lowest-power (greatest field) eyepiece, it is not too long. The resulting removable dew cap is remarkably lightweight and makes a world of difference.

For those wanting a higher-tech solution, the existing dew-shield can be actively warmed – this is the principle of the *Kendrick Dew Zapper*®. As a DIY option, many people have made perfectly serviceable heated dew caps using resistance wire or strings of resistors taped to the inside of the dew cap. These need not impinge on the light path. Those readers with electronic capabilities will, no doubt, be able to see more sophisticated solutions.

6.1.5　How can I keep dew off eyepieces?

Eyepieces offer a different problem. For obvious reasons, a long dew cap is not an option (and eyecups even make the matter worse!). The obvious thing is to avoid breathing on them, but there is another source of warm moist air: our eyes. It makes sense to dry a moist eye before putting it to an eyepiece, particularly if that eyepiece has an eyecup, which will trap any moist air. On particularly cold nights, remove the eyecup or fold it down if possible. There are two obvious ways of warming eyepieces: an inside pocket or some form of electrical heating. Of these, by far the simplest is to swap eyepieces in and out of an inside pocket.

6.1.6 What about Newtonian diagonals?

Diagonals in open-tube Newtonians can suffer from dewing. This can be reduced by a shroud at the top end, but the diagonal can sometimes still "see" a lot of sky when the telescope is used for observing very low altitude objects. In the field, one needs to be creative about warming a dewed diagonal. One solution is to cup one's ungloved hands around, but not touching, it for several minutes. It looks worse than before when the hands are removed, because the moist air from the hands has added to the condensate on the mirror, but it will normally miraculously clear after a minute or so, provided it has been sufficiently warmed. A portable hair dryer can also be used.

6.1.7 What about binoculars?

Although some older binoculars come with sliding dew-shields, these are entirely absent on "normal" binoculars made nowadays. I have made dew-caps out of cardboard poster tubes (50 mm objectives) and toilet-roll inners reinforced with gaffer tape (30 mm objectives) and, although they seem to work, the added length is awkward and they tend to get knocked when the binocular is hanging from my neck. This may be a better solution for mounted binoculars.

The solution I use nowadays is to hang the binocular inside my jacket as soon as there is any sign of dewing and, on cold nights, when I am not actually using them. If you do this, you will find that they immediately dew up even worse from the warm moist air under the jacket, but they soon clear and are ready for use. On dewy nights fold back the eyecups if you can, to allow whatever moving air there is to waft away the moist warm air from your eyes.

6.1.8 What about *Telrad*® finders?

The *Telrad*® finder is a remarkably useful accessory but, because of its exposure to the sky, the flat plate of a *Telrad*® is particularly prone to dewing. A specific solution is to use a piece of one of those "plastic and elastic" A4 document folders to make a hemi-sphericalish shield that covers the entire-length-and-possibly-more of the *Telrad*®. You can secure it with self-adhesive Velcro pads.

6.1.9 What about proprietary dew-prevention sprays?

These are of varying efficacy in preventing dew formation. They will almost certainly reduce the efficacy of anti-reflective coatings on refractive optical elements, i.e. lenses or corrector plates.

6.1.10 How do I remove dew once it has formed?

The principles of dew removal are to provide warmth and to remove the moisture. Provided that the source of warmth is a dry source, the moisture will usually quickly disappear of its own accord. Possibly the most useful dry heat source is a "traveller's" hairdryer which operates off its own batteries or from the cigarette-lighter socket of a vehicle. Proprietary "dew guns" are essentially the same things with a higher price tag. If you are using a hairdryer, do make sure that it will not propel anything abrasive (such as rucksack grit) into your expensive optic!

However, available sources of warmth are often moist (e.g. body heat under a jacket, a warm room) and, whilst the moisture will eventually disappear, this may take more time than you are willing to allow. The simplest solution is to force the moisture to evaporate by creating an airflow over it. Camera "puffer" brushes are ideal for this purpose with small optics.

6.2 Dressing for the Cold

Attending the eyepiece of the telescope will not go down in the annals of history as a physically active task; for this reason alone, we need to dress as though it was about 5° to 10° colder than it really is. This should compensate for the lack of physical activity. The principle I apply to keeping warm is one I call "sensible stratification":

6.2.1 What should I wear next to the skin?

The standard undergarment is made of cotton. If you want to keep warm, abhor it; if you wish to stick to natural fibres, change to wool (itchy) or silk (expensive), otherwise you need to go synthetic. The reason cotton is so poor is that it absorbs the moisture (that is why we use it for towels!) provided by our constant perspiration. As you may recall from school physics lessons, when moisture evaporates, it absorbs heat from its surroundings. In the case of evaporation from a cotton vest, that heat source is your chest. The solution is to wear a fabric that wicks water away from the skin without absorbing it; the best hydrophobic fabrics are made from polypropylene (yes, the same stuff that baler twine is made of!).

6.2.2 What insulation should I use?

The middle layer is the insulating layer that needs to trap as much air as it can, since air is an excellent insulator. By far the most efficient insulating layer per unit weight, or per unit volume, is dry goose down, but the damper it gets the less effective it is. It takes ages to dry out, and it is extremely expensive. The moisture that has been wicked through the inner layer comes to the middle layer, which it will dampen unless it passes through. If you wish to stick to natural fibres, wool has the reputation for being a good insulator when it is wet, although it can be a bit heavy. Modern synthetics such as *Hollofil®* and *Thinsulate®* are excellent insulators and will wick moisture away from the body without absorbing it. My favourite is a polyester material called *Polartec®*; it is a fleecy material that doesn't pill and wicks moisture away very efficiently. Even if you do manage to get it wet (unlikely to happen doing astronomy), it dries on the body very quickly.

It makes sense to have layers of garments in this middle layer, so you can adjust your insulation to different conditions. Also remember that this layer should not be too tight or be compressed by the outer layer, or its insulating properties (related to the amount of trapped air) are reduced. The adage for keeping warm is "lots of loose layers". Something with pockets is useful for keeping eyepieces warm (see Question 6.1.5).

6.2.3 And on the outside?

Whilst we tend not to observe in strong winds, even a five-knot breeze can make a great deal of difference to our comfort if it can get into the insulating layer. The outer layer needs to be wind-proof, but it must also pass the water vapour that has been wicked away from our bodies by the lower layers. Since we don't observe in the rain, there is no need to go to the expense of modern microporous waterproof fabrics like *Gore-Tex®*; something like a 3-ply *Supplex®* nylon is almost as windproof and is a better "breather".

6.2.4 What should I wear on my head?

It is said that we can lose anything up to 40% of our bodyheat through our heads. Even if this is as low as 25%, it indicates that we can regulate our body temperature by changing our headware, thereby reducing the need to fiddle about with the insulating middle layer of clothing. "Extreme conditions" headwear would follow the same pattern as our other clothing, i.e. silk or polypropylene balaclava, covered by a wool or *Polartec®* layer, covered by a windproof layer. In southern England I have never bothered with the underlayer; I find that the combinations I can get with a *Polartec®*

earband, a woollen balaclava and a *Polartec®/Supplex®* earflap hat are sufficiently versatile to enable me to keep comfortable for long periods in cold weather.

6.2.5 How can I keep my fingers warm and still use them?

There is no simple solution to the need to keep the fingers warm and also have them free and sufficiently sensitive to make fine adjustments with small knobs. A polypropylene inner glove covered by woollen finger-less gloves would offer a good combination that affords sensitivity and warmth; substituting fisherman's "slit finger" neoprene gloves for the polypropylene if you really suffer from cold fingers. Having said that, I use an old pair of leather gloves and handwarmer pockets for most of the time, substituting *Polartec®* gloves for the leather in really cold weather. The gloves come off for fiddly adjustments, such as putting a filter into an eyepiece.

6.2.6 How do I prevent my feet chilling?

On clear winter nights the ground cools faster than the air and we lose heat rapidly to the cold ground if we wear thin soles. Ordinary thick-soled shoes worn with two layers of socks (inner "wicking" sock, outer insulat-ing "cushion loop" sock) are sufficiently warm for most southern English conditions, but for the more cold-footed amongst us, or for those who live in colder climes, there are higher-tech alternatives such as "Moon Boots" that have closed-cell foam in the midsole and *Thinsulate®* in the upper. If the aesthetics bother you more than the cost, a high-altitude mountaineering boot would probably meet your requirements.

6.2.7 What exercise will warm me?

Any physical exercise will be warming. Because exer-cise increases the body metabolism for several hours, exercise shortly before an observing session is helpful.

6.2.8 What "fuel" can I use to keep warm?

One of the best ways to warm up outdoors is with a hot drink. Hot, sweet and strong fruit cordials (e.g. black-currant or orange) are probably the most effective warmers. Tea and coffee are fine (but are diuretic and cause a net loss of liquid), and hot sweet Ovaltine or chocolate would probably be better. My favourite is Tonkin's *Winter Warmer*. Caveat: sweet drinks and observing instruments are best kept apart!

In this regard, it is important to note that we lose a lot of water through our breath on crisp cold nights. The cold depresses our thirst reflex and our bodies respond to dehydration by restricting circulation to the extremities, i.e. those bits that chill most easily. For this reason it is good to have a drink immediately before an observing session.

Note that alcoholic drinks reduce night vision and increase the rate at which body heat is lost.

6.2.9 What is Tonkin's *Winter Warmer*?

Ingredients (to fill a 1 litre Thermos):

1 packet lime jelly
Apple juice concentrate
Half a cinnamon stick (or c. $\frac{1}{2}$ teaspoon cinnamon powder)
A few thin slices of fresh ginger (or c. $\frac{1}{4}$ teaspoon powdered ginger)
A pinch of nutmeg

Dissolve the jelly in boiling water. Add about 100 ml (4 fl.oz.) apple juice concentrate and make up to nearly a litre with boiling water. Adjust the taste by making up to the full litre with more apple juice or more water (make it strong!). Put the spices in the thermos flask and pour the liquid on top. Quickly put the lid on to keep the spice aromas in.

- Do drink some as soon as cold appears to be getting through to you.
- Do share it with your companions.
- Don't let it get cold in the cup (it gels!).
- Do keep it away from eyepieces and filters!

6.3 Magnitudes

6.3.1 What is magnitude?

It is the apparent brightness of a star or other object. The system now in use derives from that of Hipparchus of Nicaea in the Second Century AD. The first stars to appear at twilight were "first magnitude" and so on, down to 6th magnitude for those that were at the limit of naked eye visibility in a dark sky.

In the late 19th Century, it became apparent that the human eye has a logarithmic, not linear, response to light, and the astronomer Norman Pogson proposed a mathematically precise scale that would match that of Hipparchus. He proposed that a difference of 5 magnitudes would correspond to a 100-fold difference of brightness; i.e. a magnitude is a difference of 2.512-fold in brightness. ($2.512^5 = 100$)

Normally when amateur astronomers talk about the magnitude of a star, they are referring to its "visual" magnitude. That is the brightness of the star as can be seen by the naked eye – whether assisted or not. However, there are also magnitude scales for other parts of the electromagnetic spectrum.

6.3.2 How can the magnitude of a star be found?

The magnitude scale is calibrated by arbitrarily assigning a magnitude to a star – the magnitudes of other stars follow by comparison to this standard star. In practice, there are many standard stars, so distributed that standards of different spectral types are likely to be reasonably close to any star being observed. The magnitudes assigned to the standards are such that they are consistent with each other, and the calibration is based on a star of magnitude 1.0 having an energy of 9.87×10^{-9} W m^{-2} at the top of the Earth's atmosphere.

One should choose a star of similar spectral type when estimating a magnitude of another star – the eye and photometers are differently sensitive to different wavelengths.

Here are some magnitudes of some common objects:

- 26.8   the Sun
- 12.6   full Moon

− 4.7	Venus (at its brightest)
− 2.9	Jupiter (at its brightest)
− 2.8	Mars (at its brightest)
− 1.42	Sirius (brightest naked eye star)
0	approximate magnitude of Alpha Centauri, Arcturus, Vega, and Rigel
+ 0.7	Saturn (at opposition)
+ 1	approximate magnitude of Antares, Spica, and Pollux
+ 2	approximate magnitude of Polaris, Orion's Belt and the Big Dipper
+ 4	approximately the faintest stars visible in brightly-lit urban skies
+ 5.5	Uranus
+ 6.5	approximately the faintest stars visible in dark, transparent skies
+ 8	Neptune
+ 14	Pluto

6.3.3 What is "limiting magnitude"?

This can have different meanings according to the context.

When referred to a star atlas, it is the approximate magnitude of the faintest stars included.

When referred to the sky, it is the dimmest stars that would be visible at the zenith. It is sometimes referred to as *naked eye limiting magnitude* or NELM. It is usually estimated by counting the visible stars in a clearly defined region of sky, such as the Square of Pegasus, then referring to a standard table.

When referred to an instrument, it is the faintest stars visible with the instrument under ideal conditions. The theoretical limiting magnitude, M_{lim}, of a telescope with an aperture, D, can be calculated by the formula:

$$M_{lim} = 6.5 - 5 \log(\delta) + 5 \log(D)$$

However, the predictions of theory do not always accord with observation and the observed magnitude limits for different apertures are closer to these values for amateur instruments:

Aperture (mm)	50	100	150	200	250	300	500
M_{lim}		12	13.5	14.5	15	15.5	16	17

These are data for experienced observers in ideal conditions. Haze, inexperience, and poor equipment can all reduce the value of the limiting magnitude. See also 3.1.5, 3.1.6, and 3.1.7.

6.3.4 What is "integrated magnitude"?

It is a term applied to extended objects. The integrated magnitude is that which would apply if all the light energy from the object were coming from a point source. The use of integrated magnitudes can be misleading to visual astronomers; a common example of this is the galaxy M33, which has an integrated magnitude of about 6, leading people to believe that it should be an easy object in a small telescope or binocular. In fact, it is a large, low surface-brightness object (i.e. its light comes from a large area of sky) and it can be extremely difficult to observe until its nature is understood and you know what to look for, appearing only as a slight brightening of the sky.

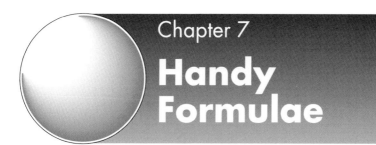

Chapter 7
Handy Formulae

7.1 Telescope Magnification

$$M = f_o/f_e$$

where f_o = focal length of objective lens or primary mirror; and f_e = focal length of eyepiece.

7.2 Exit Pupil Size

$$d = \text{aperture/magnification}$$

or

$$d = f_e/F$$

where d = exit pupil diameter; f_e = focal length of eyepiece; and F = focal ratio of objective lens or primary mirror.

7.3 Diagonal Offset

$$Offset = ma/4F$$

where ma = minor axis of diagonal mirror; and F = focal ratio of primary mirror.

7.4 Limiting Magnitude

$$M_{lim} = 6.6 - 5\log(\delta) + 5\log(D)$$

where D = aperture; and δ = diameter of pupil of eye.

7.5 Telescope Resolution

$$A = 1.22\lambda/D$$

where A = radius of Airy disc; λ = wavelength of light; and D = aperture of telescope.

7.6 Co-ordinate Conversion

$$\sin(a) = \sin(d)\sin(l) + \cos(d)\cos(l)\cos(LST - r)$$
$$\cos(A) = (\sin(d) - \sin(l)\sin(a))/(\cos(l)\cos(a))$$

where r = right ascension; LST = local sidereal time; d = declination; A = azimuth; a = altitude; and l = latitude.

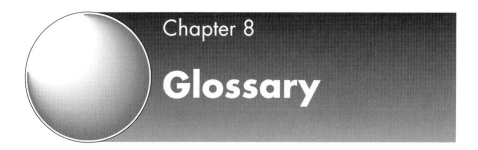

Chapter 8

Glossary

Aberration
An optical effect which degrades an image.

Absolute Magnitude
The *apparent magnitude* that an object would possess it if were placed at a distance of 10 *parsecs* from the observer. In this way, absolute magnitude provides a direct comparison of the brightness of stars.

Achromatic
Literally "no colour". A lens combination in which *chromatic aberration* is corrected by bringing two colours to the same focus.

Airy disc
The bright central part of the image of a star. It is surrounded by diffraction rings and its angular size is determined by the aperture of the telescope. About 85% of the light from the star should fall into the Airy disc.

Albedo
The proportion of incident light which a body reflects in all directions. The albedo of the Earth is 0.36, that of the Moon is 0.07 and that of Uranus is 0.93. The true albedo may vary over the surface of the object so, for practical purposes, the mean albedo is used.

Altazimuth
A mounting in which the axes of rotation are horizontal and vertical. An altazimuth mount requires motion of both axes to follow an astronomical object, but is simpler to make than an *equatorial mount* and can, in some forms, be held together by gravity.

Altitude
The angle of a body above or below the plane of the horizon – negative altitudes are below the horizon.

Anomaly
The angle at the Sun between a planet and its perihelion.

Apastron
The position in an orbit about a star at which the orbiting object is at its greatest distance from the star.

Aperture
The diameter of the objective lens or primary mirror.

Aphelion
The position in a *heliocentric* orbit at which the orbiting object is at its greatest distance from the Sun.

Apoapsis
The position in an orbit at which the orbiting object is at its greatest distance from the object about which it is orbiting.

Apochromatic
A lens combination in which *chromatic aberration* is corrected by bringing three colours to the same focus. Some manufacturers to describe achromatic doublets whose false colour is approximately equivalent to that of an apochromatic triplet lens use the term.

Apogee
The position in a *geocentric* orbit at which the orbiting object is at its greatest distance from Earth.

Apparent magnitude
The brightness of a body, as it appears to the observer, measured on a standard *magnitude* scale. It is a function of the *luminosity* and distance of the object, and the transparency of the medium through which it is observed.

Arcminute
One sixtieth of a degree.

Arcsecond
The second division of a degree of arc, equal to one sixtieth of an arcminute (i.e. 1/3600 of a degree).

Ascending node
The position in the orbit of a planet (or the Moon) where it crosses the plane of the *ecliptic*, moving northward.

Astigmatism
An optical *aberration* resulting from unequal magnification across different diameters.

Astronomical Unit (AU)
The mean distance from the Earth to the Sun, i.e. 149 597 870 km or 499.005 light seconds.

Azimuth mount
The angular distance around the horizon, usually measured from north (although it is sometimes measured from south), of the *great circle* passing through the object.

Barlow lens
A diverging lens which has the effect of increasing (usually doubling) the effective focal length of the telescope.

Bolometric magnitude
The total radiation received from an object.

Catadioptric
A telescope whose optics, not including the eyepiece, consist of both lenses and mirrors. The most common examples of these are the Schmidt-Cassegrain telescopes, whose "lens" is an aspheric corrector plate, and the Maksutov-Cassegrain telescopes, whose "lens" is a deeply curved meniscus.

Celestial Equator
The projection of the Earth's equator upon the *celestial sphere*. It is the reference plane for measurement of the *Declination*.

Celestial Sphere
The projection of space and the objects therein onto an imaginary sphere surrounding the Earth and centred on the observer.

Central meridian
The imaginary line through the poles of a planet that bisects the planetary disc.

Chromatic aberration
An *aberration* of refractive optical systems in which light is dispersed into its component colours, resulting in false colour in the image.

Collimation
The bringing of the optical components of a telescope into correct alignment.

Coma
(i)   The matter surrounding the nucleus of a comet –
      it results from the evaporation of the nucleus.
(ii) An optical *aberration* in which stellar images are
      fan-shaped, similar to comets.

Conjunction
There are at least three definitions of conjunction.
Bodies are said to be in conjunction when: (i) they
have the same *Right Ascension*; (ii) they have the same
ecliptic longitude; (iii) they are at their closest.

Culmination
An object culminates when it reaches the observer's
meridian. It is then at its greatest *altitude*.

Declination
The angle of an object above or below the *celestial
equator*. It is part of the system of equatorial co-
ordinates.

Descending Node
The position in the orbit of a planet (or the Moon)
where it crosses the plane of the *ecliptic*, moving
southward.

Diffraction limited
A measure of optical quality in which the performance
is limited only by the size of the theoretical diffracted
image of a star for a telescope of that aperture.

Dobsonian
Named after John Dobson, who originated the design.
An *altazimuth mount*, constructed usually of plywood
or MDF, suited to home construction. Also refers to a
telescope so mounted.

Eccentricity
The eccentricity of an orbit is a measure of its depar-
ture from a circle. Elliptical orbits have an eccentricity
> 0 and < 1, parabolic paths have an eccentricity $= 1$,
and hyperbolic paths have an eccentricity > 1.

Eclipse
An alignment of two bodies with the observer such that
either the nearer body prevents the light from the
further body from reaching the observer (e.g. solar
eclipse or eclipsing binary stars), or when one body
passes through Earth's shadow (e.g. lunar eclipse).

Ecliptic
The apparent path the Sun on the *celestial sphere*. It
intersects the *celestial equator* at the equinoxes. It is so

named because when the Moon is on the ecliptic solar and lunar eclipses can occur.

Elongation
The angular distance between the Sun and any other solar system body, or between a satellite and its parent planet.

Epoch
(i) The date at which a the co-ordinates on a star chart will be correct with respect to *precession*.
(ii) The date of reference in astronomical calculations.

Equation of Time
The correction which must be applied to solar time in order to obtain mean solar time.

Equatorial Mount
A mounting in which one of two mutually perpendicular axes is aligned with the Earth's axis of rotation, thus permitting an object to be tracked by rotating this axis so that it counteracts Earth's rotation.

Equinox
Literally "equal night", it refers to the time of year when day and night are of equal length. An equinox occurs when: (i) the centre of the Sun crosses the *celestial equator*; (ii) the *Declination* of the Sun is zero (i.e. when it is on the *celestial equator*).

Exit pupil
The position of the image of the objective lens or primary mirror formed by the eyepiece. It is the smallest disc through which all the collected light passes and is therefore the best position for the eye's pupil.

Eye relief
The distance from the eye lens of the eyepiece to the *exit pupil*. Spectacle wearers require sufficient eye relief to enable them to place the eye at the exit pupil.

Eye ring
An alternative name for the *exit pupil*.

Faculae
Unusually bright spots on the Sun's surface.

First Point of Aries (FPA)
The Vernal *Equinox* point, i.e. that where the centre of the Sun, moving northwards, crosses the equator. It is the reference direction for the equatorial system of co-ordinates.

Focal plane
The plane (usually this is actually the surface of a sphere of large radius) where the image is formed by the main optics of the telescope. The eyepiece examines this image.

Focuser
The part of the telescope which varies the optical distance between the objective lens or primary mirror and the eyepiece. This is usually achieved by moving the eyepiece in a drawtube, but in some catadioptric telescopes it is the primary mirror that is moved.

Fork mount
A mount where the telescope swings in declination or in altitude between two arms. It is suited only to short telescope tubes, such as Cassegrains and variations thereof. It requires a *wedge* to be used equatorially.

Galilean Moons
The four Jovian moons first observed by Galileo (Io, Europa, Ganymede and Callisto). They are observable with small amateur telescopes.

Geocentric
Earth-centred.

Geostationary orbit
The orbit of a satellite which is both *geosynchronous* and in the equatorial plane. The satellite will appear to remain in a fixed position in relation to the observer.

Geosynchronous orbit
The orbit of a satellite in which the orbital period of the satellite is equal to Earth's period of rotation. If the orbit is in the equatorial plane, the satellite will be *geostationary*; if the orbit is inclined to the equatorial plane the satellite will appear to trace a lemniscate in the sky.

German Equatorial Mount (GEM)
A common *equatorial mount* for small and medium sized amateur telescopes, suited to both long and short telescope tubes. The telescope tube is connected to the counter-weighted declination axis, which rotates in a housing that keeps it orthogonal to the polar axis. Tracking an object across the meridian requires that the telescope be moved from one side of the mount to the other, which in turn requires that both axes are rotated through 180°, thus reversing the orientation of the image. This is not a problem for visual observation, but is a limitation for astrophotography.

Granulation
The "grains of rice" appearance of the Sun's surface, which results from convection cells within the Sun.

Great circle
A circle formed on a the surface of a sphere which is formed by the intersection of a plane which passes through the centre of a sphere. A great circle path is the shortest distance between two points on a spherical surface.

Heliocentric
Sun-centred.

Hour Angle
The angle, measured westwards around the *celestial equator*, between the observer's *meridian* and the *hour circle* of an object.

Hour Circle
(i) A *great circle* passing through an object and the celestial poles.
(ii) The setting circle on the polar axis of an *equatorial mount*.

Inclination
(i) The angle between the *ecliptic* and the orbital plane of a planet.
(ii) The angle between the orbital plane of a satellite and the equatorial plane of the parent body.

Inferior Conjunction
The *conjunction* of Mercury or Venus when they lie between the Earth and the Sun.

Inferior Planets
Planets (i.e. Mercury and Venus) whose orbits lie inside the Earth's orbit.

Integrated Magnitude
The magnitude which would apply if all the light energy from an extended object was coming from a point source.

Light bucket
Slang term for a telescope of large aperture.

Light Year
The distance travelled by light in one year: 9.4607×10^{12} km, or 63 240 AU, or 0.3066 *parsecs*.

Limb
The edge of the disc of a celestial body.

Luminosity
The amount of energy radiated into space per second by a star. The bolometric luminosity is the total amount of radiation at all frequencies; sometimes luminosity is given for a specific band of frequencies (e.g. the visual band).

Magnitude
The brightness of a celestial body on a numerical scale. See also *absolute magnitude, apparent magnitude, bolometric magnitude* and *integrated magnitude.*

Mean anomaly
The *anomaly* which would exist if a planet orbited at a uniform speed in a circular orbit.

Meridian
The *great circle* passing through the celestial poles and the observer's *zenith.*

Meteor
The incandescent trail of a *meteoroid* as it enters the Earth's atmosphere.

Meteorite
A *meteoroid* which reaches Earth's surface.

Meteoroid
A fragment of matter which may turn into a *meteor* or a *meteorite* if it strikes the Earth.

Minor Planets
Another term for asteroids.

Nadir
The point on the *celestial sphere* directly below the observer. Opposite of *zenith.*

Obliquity of the Ecliptic
The angle between the plane of the *ecliptic* and that of the *celestial equator.*

Occultation
An alignment of two bodies with the observer such that the nearer body prevents the light from the further body from reaching the observer. The nearer body is said to occult the further body. A solar *eclipse* is an example of an occultation.

Opposition
The position of a planet such that the Earth lies between the planet and the Sun. Planets at opposition are closest to Earth at opposition and thus opposition offers the best opportunity for observation.

OTA
Abbreviation for Optical Tube Assembly. It is normally considered to consist of the tube itself, the focuser and the optical train from the objective lens (refractor), primary mirror (reflector), or corrector plate (catadioptrics) up to, but not including, the eyepiece.

Parsec
The distance at which a star would have a parallax of one arcsecond. (3.2616 *light years*, 206 265 *astronomical units*, 30.857×10^{12} m).

Penumbra
Literally "next to the *umbra*".
(i) The shadow that results when only part of the bright object is occulted; e.g. an observer will see a partial *eclipse* when he is in the penumbra of the shadow of the moon.
(ii) The lighter area surrounding a sunspot.

Periapsis
The position in an orbit at which the orbiting object is at its least distance from the object about which it is orbiting.

Periastron
The position in an orbit about a star at which the orbiting object is at its least distance from the star.

Perigee
The position in a *geocentric* orbit at which the orbiting object is at its least distance from Earth.

Perihelion
The position in a *heliocentric* orbit at which the orbiting object is at its least distance from the Sun.

Phase
The percentage illumination, from the observer's perspective, of an object (normally a planet or the Moon).

Planisphere
The projection of a sphere (or part thereof) onto a plane. It commonly refers to a simple device which consists of a pair of concentric discs, one of which has part of the *celestial sphere* projected onto it, the other of which has a window representing the horizon. Scales about the perimeters of the disk allow it to be set to show the sky at specific times and dates, enabling its use as a simple and convenient aid to location of objects.

Precession
A rotation of the direction of the axis of rotation. Normally refers to the precession of the equinoxes, a consequence of the effect of the Sun's gravity on the Earth's equatorial bulge. Earth's axis of rotation precesses with a period of about 25 800 years, during which time the equinoxes make a complete revolution about the *celestial equator*. Because the Vernal *Equinox* is the reference direction for the equatorial co-ordinate system, the co-ordinates of "fixed" objects changes with time and must therefore be referred to an *epoch* at which they are correct.

Prime meridian
The polar great semi-circle adopted as the reference direction for measurement of longitude. The Earth's prime meridian is the Greenwich meridian.

Prograde
The apparent eastward motion of a planet with respect to the stars.

Proper motion
The apparent motion of a star with respect to its surroundings.

Quadrature
The position of a body (Moon or planet) such that the Sun-body-Earth angle is 90°. The *phase* of the body will be 50%.

Radiant
The position in the sky from which a meteor shower appears to radiate.

Rayleigh criterion (Rayleigh limit)
Lord Rayleigh, a 19th Century physicist, showed that a telescope optic would be indistinguishable from a theoretical perfect optic if the light deviated from the ideal condition by no more than one quarter of its wavelength.

Red shift
The lengthening of the wavelength of electromagnetic radiation resulting from one or more of three causes:
Doppler red shift: resulting from bodies moving away from each other in space.
Gravitational red shift: resulting from strong gravitational fields.
Cosmological red shift: resulting from the expansion of space-time itself.

Reflector
A telescope whose optics, apart from the eyepiece, consist of mirrors.

Refractor
A telescope whose optics consist entirely of lenses.

Resolution
A measure of the degree of detail visible in an image. It is normally measured in arcseconds.

Retrograde
Apparent westward movement of a planet with respect to the stars.

Right Ascension (RA)
The angle, measured eastward on the *celestial equator*, between the *First Point of Aries* and the *hour circle* through the object.

Scintillation
The twinkling of stars, resulting from atmospheric disturbance.

Semi-major axis
Half the distance across an ellipse measured through its foci.

Sidereal Time
The *hour angle* of the *First Point of Aries*. It is time measured with respect to the stars.

Solstice
Literally "sun still". It refers to the apparent standstill of sunrise and sunset points at midsummer and midwinter.
(i) The most southerly and northerly declinations of the Sun.
(ii) The date on which the Sun attains its greatest *Declination*.

Spherical aberration
An optical *aberration* in which light from different parts of a mirror or lens is brought to different foci.

Superior conjunction
The *conjunction* of Venus and Mercury when they are more distant than the Sun.

Superior Planets
Those planets whose orbits lie outside the Earth's orbit.

Terminator
The boundary of the illuminated part of the disc of a planet or moon.

Transit
(i)   The passage of Mercury or Venus across the disc of the Sun.
(ii) The passage of a planet's moon across the disc of the parent planet.
(iii) The passage of a planetary feature (such as Jupiter's Great Red Spot) across the *central meridian* of the planet.
(iv) The passage of an object across the observer's *meridian* (see also *culmination*).

Umbra
(i)   The shadow that results when a bright object is completely occulted. A total *eclipse* of the Sun occurs when the observer is in the Moon's umbra.
(ii) The dark inner region of a sunspot.

Wedge
The part that fits between the tripod or pillar and the fork of a fork-mounted telescope, which enables the fork to be equatorially aligned.

Worm drive
Probably the most common drive on equatorial mounts. It consists of a spirally cut cylinder (the "worm") which rotates longitudinally such that its thread engages with the specially shaped teeth on the circumference of a disc (the "worm wheel"), which in turn drives the shaft of the mount.

Zenith
The point on the *meridian* directly above an observer.

Zenithal Hourly Rate (ZHR)
The theoretical hourly rate of meteors which would be observed at the peak of a shower, by an experienced observer, with the *radiant* at the *zenith*, under skies with a limiting naked eye *magnitude* of 6.5.

Chapter 9

Bibliography

Berry R., *Choosing and Using a CCD Camera*, 1992, Richmond, VA: Willmann-Bell; ISBN 0-943396-39-5.
Excellent guide.

Covington M., *Astrophotography for the Amateur*, 1999, Cambridge: Cambridge University Press; ISBN 0-521-62740-0.
Comprehensive overview, suitable for beginner and intermediate level.

Duffett-Smith P., *Practical Astronomy with your Calculator*, 1988, Cambridge: Cambridge University Press; ISBN 0-521-35699-7.
Comprehensive and useful, with copious worked examples.

Harrington P., *Star Ware*, 1998, New York: John Wiley and Sons; ISBN 0-471-18311-3.
Good guide to choosing equipment.

Kitchin C.R., *Telescopes and Techniques*, 1995, London: Springer-Verlag; ISBN 3-540-19898-9.
Very good theoretical introduction top practical astronomy.

Meeus J., *Astronomical Algorithms*, 1991, Richmond, VA: Willmann-Bell; ISBN 0-943396-35-2.
In depth coverage. Excellent.

Moore P., *Exploring the Night Sky with Binoculars*, 1986, Cambridge: Cambridge University Press; ISBN 0-521-36866-9.
All you need to get going in binocular astronomy.

North G., First Steps in Astronomical Calculations, in Moore P., *The Modern Amateur Astronomer*, 1995, London: Springer-Verlag; ISBN 3-540-19900-4.
Good introduction.

Suiter H.R., *Star Testing Astronomical Telescopes, A Manual for Optical Evaluation and Adjustment*, 1994, Richmond, VA: Willmann-Bell; ISBN 0-943396-44-1.
The acknowledged bible on star-testing.

Star atlases

Ridpath I., *Norton's Star Atlas and Reference Handbook*, Harlow: Addison Wesley Longman; ISBN 0-582-31283-3.
More than just an atlas – the reference handbook section, which makes up most of the bulk, is excellent.

Ridpath I. and Tirion W., *Collins Gem Stars*, 1999, Glasgow: Harper-Collins; ISBN 0-00-472474-7.
The portable star atlas – it fits in a shirt pocket!

Tirion W., *Sky Atlas 2000*, 1999, Cambridge: Cambridge University Press; ISBN 0-521-65431-9.
Goes to magnitude 8.5. This ISBN is the Field Edition – other editions are available.

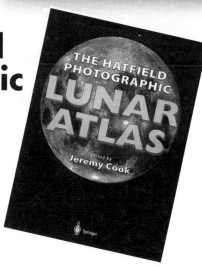